# Information Warfare in Business

*Information Warfare in Business* provides a significant and highly topical perspective on the concept of the network organization. Information warfare covers a diverse range of techniques that are transforming both military affairs and business practice. Business and military organizations are increasingly reliant upon information technology, which has become a valuable resource and a weapon in its own right. Individuals and organizations are now under threat from information warfare attacks and must learn to guard against them.

This book outlines the many forms that information warfare can take and the various weapons at its disposal. These include high technology approaches such as hacking, identity theft and the spread of computer viruses, and low technology approaches such as the use of misinformation and propaganda and the execution of confidence tricks to get hold of restricted information. Over recent years non-business organizations, such as Napster and Linux, have emerged that have developed nomadic strategies to pursue their aims and resist the power of global multinational corporations. Similar strategies have also been employed on a political level by protest groups like the Zapatista rebels in Mexico.

The issue of power is central to the general argument of this book. Information warfare has emerged as part of a gradual movement away from disciplinary forms of organization, towards a society organized around free-floating networks of control. The means of production of the new service economies are increasingly orientated around the exploitation of the employees' minds, their emotions and their interactions with others. These same tools and resources have been transformed into the weapons of information warfare. Iain Munro's underlying theoretical analysis is deeply indebted to the social theories of Michel Foucault, Paul Virilio, Antonio Negri, Gilles Deleuze and Felix Guattari.

This multidisciplinary text should be of interest to anyone studying the role of networks in society, whether in management science, business studies, organizational theory and behaviour, international relations, information systems or sociology.

**Iain Munro** is a Lecturer at the Department of Management, University of St Andrews. He has published research in the fields of organization studies, operational research and business ethics.

## Routledge studies in business organizations and networks



1 **Democracy and Efficiency in the Economic Enterprise**
*Edited by Ugo Pagano and Robert Rowthorn*

2 **Towards a Competence Theory of the Firm**
*Edited by Nicolai J. Foss and Christian Knudsen*

3 **Uncertainty and Economic Evolution**
Essays in honour of Armen A. Alchian
*Edited by John R. Lott Jr*

4 **The End of the Professions?**
The restructuring of professional work
*Edited by Jane Broadbent, Michael Dietrich and Jennifer Roberts*

5 **Shopfloor Matters**
Labor-management relations in twentieth-century American manufacturing
*David Fairris*

6 **The Organisation of the Firm**
International business perspectives
*Edited by Ram Mudambi and Martin Ricketts*

7 **Organizing Industrial Activities Across Firm Boundaries**
*Anna Dubois*

8 **Economic Organisation, Capabilities and Coordination**
*Edited by Nicolai Foss and Brian J. Loasby*

9 **The Changing Boundaries of the Firm**
Explaining evolving inter-firm relations
*Edited by Massimo G. Colombo*

10 **Authority and Control in Modern Industry**
Theoretical and empirical perspectives
*Edited by Paul L. Robertson*

11 **Interfirm Networks**
Organization and industrial competitiveness
*Edited by Anna Grandori*

12 **Privatization and Supply Chain Management**
*Andrew Cox, Lisa Harris and David Parker*

13 **The Governance of Large Technical Systems**
*Edited by Olivier Coutard*

14 **Stability and Change in High-Tech Enterprises**
Organisational practices and routines
*Neil Costello*

15 **The New Mutualism in Public Policy**
*Johnston Birchall*

16 **An Econometric Analysis of the Real Estate Market and Investment**
*Peijie Wang*

17 **Managing Buyer-Supplier Relations**
The winning edge through specification management
*Rajesh Nellore*

18 **Supply Chains, Markets and Power**
Mapping buyer and supplier power regimes
*Andrew Cox, Paul Ireland, Chris Lonsdale, Joe Sanderson and Glyn Watson*

19 **Managing Professional Identities**
Knowledge, performativity, and the 'new' professional
*Edited by Mike Dent and Stephen Whitehead*

20 **A Comparison of Small & Medium Enterprises in Europe and in the USA**
*Solomon Karmel and Justin Bryon*

21 **Workaholism in Organizations**
Antecedents and consequences
*Ronald J. Burke*

22 **The Construction Industry**
An international comparison
*Edited by Gerhard Bosch and Peter Philips*

23 **Economic Geography of Higher Education**
Knowledge, infrastructure and learning regions
*Edited by Roel Rutten, Frans Boekema and Elsa Kuijpers*

24 **Economies of Network Industries**
*Hans-Werner Gottinger*

25 **The Corporation**
Investment, mergers and growth
*Dennis C. Mueller*

26 **Industrial and Labour Market Policy and Performance**
Issues and perspectives
*Edited by Dan Coffey and Carole Thornley*

27 **Organization and Identity**
*Edited by Alison Linstead and Stephen Linstead*

**28 Thinking Organization**
*Edited by Stephen Linstead and Alison Linstead*

**29 Information Warfare in Business**
Strategies of control and resistance in the network society
*Iain Munro*

**30 Business Clusters**
An international perspective
*Martin Perry*

# Information Warfare in Business

## Strategies of control and resistance in the network society

**Iain Munro**

Routledge
Taylor & Francis Group
LONDON AND NEW YORK

First published 2005
by Routledge
2 Park Square, Milton Park, Abingdon, Oxon OX14 4RN

Simultaneously published in the USA and Canada
by Routledge
270 Madison Ave, New York, NY 10016

*Routledge is an imprint of the Taylor & Francis Group*



Typeset in Times by Wearset Ltd, Boldon, Tyne and Wear
Printed and bound in Great Britain by MPG Books Ltd, Bodmin



*British Library Cataloguing in Publication Data*
A catalogue record for this book is available from the British Library

*Library of Congress Cataloging in Publication Data*
A catalog record for this book has been requested

ISBN 0-415-33968-5

**This book is dedicated to Kyle, Stan, Kenny and Eric**

# Contents

# Acknowledgements

I would like to acknowledge the help and support of all my friends and colleagues for their invaluable conversations that have helped me develop the ideas presented in this book, most particularly Duncan, John, Chris, Steffen and Gibson.

# 1 Information warfare in business

Every man is surrounded by a neighbourhood of voluntary spies.

(Jane Austen, *Northanger Abbey*)

Power is impenetrable. The man who has it sees through other men, but does not allow them to see through him.

(Elias Canetti, *Crowds and Power*)

This book describes the emergence of information warfare and the distinctive role that such warfare is coming to play in both business organizations and in society in general. A number of other related themes will be discussed alongside the concept of information warfare, including strategy, power, the network society, knowledge management, and identity. There have already been a number of excellent commentaries on the way in which the Internet has transformed business, for example Manuel Castells's work on the network society, and Michael Porter's later work on the influence of the Internet on corporate strategy (Castells 1996, Porter 2001). This book is distinctive in showing the important role of information warfare in such developments. Information warfare includes a diverse range of practices, all of which concern the security of our information resources, what we know about ourselves and what we know about the world around us. This includes both high-technology and low-technology techniques, from informal networks of gossip and rumour, to the more systematic approaches of public relations and propaganda, to entirely technological techniques of hacking and the creation of computer viruses.

The general theoretical approach of the work is sympathetic to Castells' idea of the network organization and its association with the breakdown of traditional hierarchical social relations (Castells 1996). However the argument also draws upon Deleuze and Guattari's work (1988), by contrasting nomadic social movements with sedentary forms of organization. The network organization can adopt a sedentary form as a controlling mechanism that captures information and regulates ideas, or as a nomadic free flowing set of relations. This distinction provides a general point of differentiation from existing works on the network organization which tend to focus primarily on attempts to regulate networks, especially as found in the literature on knowledge management. This approach to the study of organizations highlights the 'nomadic' network as both a technological assemblage and a social formation. Today, the business world is being transformed by non-business organizations, such as the Linux and Napster user communities, which have defied the traditional models of competition. These organizations have appeared from out of the blue, posing a serious threat to their business rivals, and yet they are not themselves driven by the profit motive. They are driven instead by goals and strategies that are closer to the strategies of a guerrilla band and a more nomadic way of thinking. The theory of information warfare developed here explains the emergence of strategies in both business organizations and the military world based around the new communication technologies. This theory also explains the forces underlying a movement away from traditional disciplinary mechanisms of power, where disciplinary power is being supplemented and supplanted by more abstract networks of control. Before proceeding to a more detailed introduction to the chief arguments of this book, a few vignettes will be described, each of which illuminates the rise of information warfare.

In 1946, the Marx Brothers released *A Night In Casablanca*, a parody of wartime romances in the vein of the classic 1942 drama *Casablanca*. The Warner Brothers film studios responded by threatening to sue the Marx Brothers for infringement of their rights to the *Casablanca* brand name. Groucho handled the threatened legal action in a suitably comic manner, and wrote back to the film studio stating that they might as well try suing for the rights to use the name 'Brothers', which was another piece of intellectual property that the studio happened to share with the comic trio. In his letter, Groucho argued that, 'Professionally, we were brothers long before you were.... And even before us there had been other brothers...' (Marx 1948).

Understanding the legal quagmire that they might end up in, the Warner Brothers did not pursue their complaint any further, and the Marx Brothers got their way. *The Guardian* columnist John Naughton used Groucho's story to illustrate some of the absurdities that surround today's wrangles over intellectual property rights. What was, in 1946, an absurd exchange between a comic genius and a powerful film studio is fast becoming the norm in a society where the circulation of intellectual property forms a keystone of the economy. For example, in 2002, Microsoft filed a complaint against Lindows.com, an organization that distributes a Linux-based operating system free on the Internet. Microsoft lawyers believed that the word Lindows was an imitation of the 'Windows' brand name and hence an infringement of their intellectual property. Microsoft was concerned that the Lindows package may eat into its market share because

Linux was already quite popular, has a graphical user interface, and is free. Lindows was considered good enough to be pre-installed on the cheap PCs being sold by the retail giant Wal-Mart in the US. There is a clear parallel between the Microsoft/Lindows debacle and the Warner Brothers/Marx Brothers case, where Microsoft was acting in a proprietary manner in regard to using the word 'windows', even though there had been many windows of one form or another well before Microsoft had developed its own software.

The Lindows example is by no means an isolated incident, and the battle between Microsoft and Linux is long standing. Many of the computer viruses that circulate the Internet are targeted at Microsoft programmes specifically, and the users of Linux software have been largely safe from attack or infection. Microsoft has even devoted a website to anti-Linux propaganda, or what it terms 'Linux myths', claiming that Linux is unreliable and compares unfavourably to its own Windows software. Despite this, Linux usage is on the increase. When understood as a whole, this strategy to undermine Linux using public relations, propaganda and intellectual property laws, may be seen as a general campaign of information warfare within the business community. The gang of computer programmers who created and developed Linux might themselves be compared with a loose-knit band of guerrilla fighters, irregulars dedicated to a clear set of ideals, not least being the free exchange of information across the Internet.

Naomi Klein's book *No Logo* contains a host of similar examples, which she describes as 'copyright bullies', where large corporations have flexed their legal muscles to ward off any potential threats to their brand name or intellectual property. Klein concedes that artists and inventors really do need to be protected from thievery, reproaching anti-copyright radicals as being provocative rather than practical. That said, Klein goes on to give a list of instances that she believes demonstrates rather too much enthusiasm on the part of large corporations to appeal to copyright laws to restrict our freedom of expression. A classic example of this corporate heavy handedness occurred when the toy company Mattel used copyright laws to block the release of a movie. The film in question was Todd Haynes', *Superstar: The Karen Carpenter Story*, based on the life story of the anorexic pop star. Mattel could interfere in the release of this movie because its director had taken the unusual step of using a cast entirely peopled by Barbie dolls, which happens to be one of Mattel's best-selling products. Although this is a somewhat unusual example, Klein is genuinely concerned that when we communicate using the language of brands and corporate logos we run the risk of being sued. The legal restrictions placed on the ownership of ideas and the circulation of intellectual property is a much wider concern, as has been pointed out by the Nobel Prize winning economist Joseph Stiglitz (2002). As a former economist for the World Bank, his work is quite critical of TRIPS, the international agreement on Trade-Related Aspects of Intellectual Property, and he has questioned the wisdom of such restrictions where they have resulted in the prosecution of companies that are seemingly engaged in projects of great social benefit, for example producing cheap generic drugs for the treatment of AIDS in impoverished countries. Even though TRIPS was not purposively designed to do people harm, much human misery has certainly resulted from such tight protection of intellectual property. In this case, the close protection of information resources by the wealthy could easily be interpreted as a kind of information warfare against the impoverished. One need not overtly attack someone to do them harm; one need only withhold a vital resource.

During the 1990s, Kevin Mitnick became notorious as the most wanted hacker in the United States. He went under several false identities to avoid capture including: Eric Weiss, a legal assistant, Brian Merrill, a hospital assistant, Houdini, a notorious hacker, and the real Kevin Mitnick. Mitnick's recent legal battle reads like it was taken from the plot of a second-rate science fiction novel, rather than being a genuine ground-breaking legal case. Mitnick is one of a select group of US citizens who have been legally prohibited from using or possessing any type of information technology, including a mobile phone or an Internet connection. As part of his legal defence his lawyers argued that he had been diagnosed as suffering from the new psychological disorder of 'computer addiction'. His crime, according to the FBI and companies like Sun Microsystems, was the theft of important data costing millions of dollars. However, Mitnick never actually made any money from the systems he infiltrated, nor did he cause them any direct damage. In Mitnick's own eyes, the worse thing he did was to charge other people for his use of the telecommunications networks. In effect, his crime was that of looking at material he had not been authorized to look at. Despite this, he ended up serving a five-year prison sentence. He served some of this period in solitary confinement, imposed on him in the belief that he had the ability to inflict massive damage to US communication networks, even whilst safely ensconced behind the walls of a prison. It was even rumoured that Mitnick possessed the ability to launch a nuclear missile by means of his extraordinary hacking skills. Perhaps his skills and his threat to society were exaggerated, nevertheless he was considered sufficiently dangerous to be placed into solitary, utterly cut off from the world outside. Since his release from prison, Mitnick has worked as an information security consultant and is the author of several books on this subject. He is, however, forbidden from writing about the specific misdeeds for which he ended up in prison. Since his release, Mitnick's work has emphasized the fact that one need not resort to sophisticated techniques of computer hacking to extract valuable information from a person or corporation. No matter what technological precautions are taken to increase the security of information, humans are by nature social beings and are therefore always vulnerable to leaking vital information. In Mitnick's own words, 'Every computer system in the world has at least one human that uses it. So, if the attacker is able to manipulate people who use the systems, the obscurity of the system is irrelevant' (Mitnick 2002: 82). The main skill required for this approach to hacking is not so much any expertise in computer programming, as it is the ability to quickly build trust, to mislead and then to execute a confidence trick. Despite Mitnick's own technological mastery, he prefers to emphasize the social elements of hacking, where technology is only one aspect of security within a wider network of social interaction. In many ways, his case is exceptional but it also provides an exemplar of the kinds of underlying social changes that are underway concerning the value placed on information, the importance of intellectual property and the regulation of access to the global flows of knowledge and information. Precisely the same trends in the technological, social and economic infrastructure have given rise to the emerging doctrine of information warfare.

Intelligence agencies, such as the CIA and the NSA, devote massive resources to protecting their computer systems from the outside. But no matter what their technological precautions, they are always vulnerable to the human factor. This was no more apparent than in the case of John Deutch, the former director of the CIA, who faced criminal charges after using an unauthorized and insecure computer for dealing with classified intelligence information (Powers 2002). As it turned out, the CIA's investigators found no evidence that his computer has actually been compromised. However, they did find evidence that this insecure computer had also been used to access pornographic material on the Internet. Visitors to such sites can sometimes leave their computers open to subsequent attacks from hackers and to a host of computer viruses. 'Troll' programs have been developed by hackers, which install themselves on the computers of visitors to these sites, and pass information from the infected computer back to the hacker. Despite this apparently serious breach of security by John Deutch, who was then a senior member of the CIA, he was later pardoned by President Clinton and the legal case against him was dropped. Information warfare is being taken very seriously by our intelligence services today. This is no more apparent when looking at the current director of the National Security Agency, Lieutenant General Michael V. Hayden, who previously held a job investigating information warfare for the Joint Chiefs of Staff.

The futurists Alvin and Barbara Toffler have observed that, 'Throughout history, the way men and women make war has reflected the way they work' (Toffler and Toffler 1993: 35). In ancient times, war was normally a seasonal activity since the abundance of food required to feed soldiers was available only at certain times of the year. Likewise, the object of warfare tended to be the capture or destruction of the enemy's agriculture. The logistics, communications and weapons of warfare tend to mirror the tools that we use for work. With the mechanization of work came the mechanization of warfare. Now that the productive basis of society is becoming increasing orientated around information, so are the weapons of warfare. Paul Virilio (2000a) has remarked that the United States has become a pioneer in the development of a military doctrine for information warfare during the first Gulf War and the Balkans conflict. According to Virilio, the power of this new form of warfare 'rests on three fundamental principles: the permanent presence of satellites over territories, the real-time transmission of the information gathered and, lastly, the ability to perform rapid analysis of the data transmitted to the various general staffs' (Virilio 2000a: 18). It is worth noting that these three fundamental principles of information warfare described by Virilio are also revolutionizing the way we do business and the wider global flows of commodities and finance.

Literature also has much to teach us about the techniques and methods of information warfare. The writer William S. Burroughs is one of the most insightful and original voices on information warfare in his literary experiments and in his essays on methods of social control (Burroughs 1986, Burroughs and Odier 1989). Burroughs conceived of a dystopia based around the control mechanisms of drug addiction and the mass media. The desire for control is itself something that one can become addicted to. His work is highly critical of the monopoly power of the media, which he saw as a real threat to the democratic process. Burroughs gave his own distinctive approach to the subversion of these mechanisms of control the name 'civilian defense'; nevertheless, the methods he proposed are almost identical to the definitions of information warfare given by leading exponents in computer security (Denning 1999) and military strategy (Arquilla and Rondfeldt 1997). He conceived of his dystopia as built up of secret organizations, some being official arms of the state apparatus, such as the intelligence services, and others being subversive organizations that hope to hide their activities from the prying eyes of the official apparatus. Each of these groups uses information and the power of communications technology as their preferred weapon of choice. Burroughs himself was clearly on the side of the subversives, giving his fictional rebel societies names like 'Civilian Defense' and the 'Johnson Family'. According to Burroughs, the monopoly of the mass media constitutes a huge machine for laying down lines of association in our subconscious, telling us what to buy, who to vote for, and how to live. His books are full of techniques for scrambling the lines of association that the mass media builds in our subconscious, and of lessons in the disruption of its propaganda. He even coined a slogan for this approach to civil resistance, 'You have nothing to lose except your prerecordings' (Burroughs 1968: 129). The citizen is a guerrilla fighter engaged in an information war against forms of monopoly control that are exercised by the massive corporate conglomerates. Burroughs recommended that citizens make use of their own recording devices and communications networks for disseminating their alternative stories and viewpoints. He noted that the electronics revolution was providing many more opportunities for people to take the media into their own hands (Burroughs and Odier 1989). With the emergence of the Internet, these channels for communication have expanded even further and the kinds of information warfare that have been developed within these networks is precisely the kind of disruptive activity that was anticipated by Burroughs.

The practice of information warfare is appearing throughout many diverse areas of society, performed by isolated individuals or as part of business strategy, as part of national security or as part of military strategy. The main themes of the book are divided into three parts: (i) creative forms of organization, (ii) the use of information as a weapon in the network society, and (iii) the power relations associated with these changes. A brief description of the scope and purpose of each part now follows.

## Part I: Creative organizational forms

The first part of the book comprises two chapters that examine creative organizational forms that have arisen within the network society. Chapter 2 describes the emergence of organizations that appear to have adopted principles remarkably similar to those of guerrilla warfare in their fight for survival. These organizations tend to use information technology as a cornerstone of their strategy, which allows them to exist as distributed networks, either as individuals or as cells. A number of strategic principles are derived from T.E. Lawrence's approach to guerrilla warfare. These principles revolve around the importance of ideas, communication and speed in overcoming one's opponents. The implementation of these principles can be greatly facilitated using advances in information technology, and as such they are particularly well suited to the concept of information warfare. A variety of different organizations are shown to have successfully used similar principles in their growth, including the Zapatista rebels in Mexico, the Linux and Napster Internet communities and the international Slow Food Society. Taken-for-granted assumptions, such as the pursuit of wealth or the expansion of territories, are challenged by the nomadic strategies that have been developed by such groups. In this respect, these principles may also be seen as a general strategy for groups that are attempting to resist some of the effects of globalization; however, the innovations made by these organizations are often also adopted and exploited by business organizations.

Chapter 3 explores recent innovations in the treatment of knowledge in organizations, most notably the invention of Intellectual Property Rights and business fads such as Knowledge Management. These approaches for capturing and domesticating knowledge are leading to intensified levels of surveillance over an increasing traffic of information, both for economic and judicial reasons. One of the main limitations in monitoring and mapping flows of knowledge and information is that much of it is actually invisible, hidden as the thoughts in workers' brains and in the tacit skills of their bodies. Approaches to knowledge management that focus mainly on techniques for mapping knowledge fail to capture the hidden tacit aspects of the creative process. Therefore, this book highlights the ways in which knowledge remains undomesticated and proposes a creative or nomadic philosophy of knowledge production. This argument will contrast those tendencies towards the increasing commodification of knowledge with counter currents such as the call for an 'intellectual commons'.

## Part II: The weapon of information

Part II outlines the idea that information is increasingly being used as a weapon, both in the civilian and the military worlds. The virus is the paradigmatic form of information warfare on the Internet; however, information warfare embraces a range of other disruptive weapons including propaganda, intelligence and the idea of speed itself – the ability to spring up anywhere. Researchers for the RAND Corporation have observed that the distributed network appears to be the ideal form of organization for practitioners of information warfare (Arquilla and Rondfeldt 1997). The network is not only the ideal structure for groups that practice information warfare, in its technological embodiment in the Internet it is also both an important means of delivering an attack and an object of attack itself.

Chapter 4 discusses different concepts of information warfare, and its relation to business, international relations, and society. A number of different concepts have been developed about the nature of information warfare that can be divided into different camps. For example, the RAND Corporation is currently investing millions of dollars of research into information warfare to protect the integrity of the US economic and military infrastructure. Researchers from the RAND Corporation propose that information can be used as a 'force multiplier' in the execution of military operations (Arquilla and Rondfeldt 1997). In contrast to this approach, the studies of information warfare undertaken by Paul Virilio take a more critical stance to the blurring of boundaries between information warfare, economic conflict and open military clashes. Particular attention will be paid to the Zapatista rebels in Mexico, who have been pioneers in the tools of information warfare and have exploited both social and technological networks of communication to great advantage. This group has already been discussed by the likes of Manuel Castells and researchers at the RAND Corporation for being exemplary practitioners of information warfare. There are also a number of thinkers in areas such as propaganda, marketing, terrorism and politics who do not explicitly use the term 'information warfare' themselves but whose work is very important in the ongoing development of this concept.

Chapter 5 examines the effects of information warfare and communication networks on the experience of subjectivity. The virus is a key weapon in the armoury of information warfare, but the implications of the virus metaphor are more far-reaching. The virus mechanism can be seen as a vehicle for the transmission of disease and harmful computer code, but it can also be used as a metaphor for the way in which ideas, information, and languages also spread throughout the globe. As such, it is a powerful metaphor for describing the transformation of social relations in the network society. This viral transmission of information and ideas is used to explain the phenomenon of cultural and linguistic homogenization within the globalizing networks of communication and trade. Another major consequence of the intensification of networks of communication and production has been a serious concern for the security of these networks and their protection against contamination from unwanted intrusion and infection. The power of the virus represents a challenge to the integrity of organizations and the maintenance of their boundaries. This analysis may be seen as a contribution to the discussion of organizational metaphors that was originally outlined in the work of Gareth Morgan. His *Images of Organization* proposed a series of perspectives from which organizations can be viewed, each based on a different root metaphor, including as a machine, as an organism, as a culture, as a brain, as a psychic prison, as a political system, as a system of domination, and as a system of flux. The significance of communication networks was discussed only briefly in Morgan's work, but it might prove fruitful to add the network metaphor to the list he provides, alongside the virus that follows in its shadow.

## Part III: Changing power relations

Part III examines how the underlying power relations in society are changing with the advent of the information-age. In particular this part shows how some of the traditional disciplinary mechanisms of power, which have been described by Michel Foucault, are being supplemented, or indeed replaced, by more invasive network forms of power. Particular attention is devoted to an analysis of the increasing importance of 'real-time' control technologies in the post-modern workplace. This part also examines the ways in which those subject to these mechanisms of control attempt to subvert their effects.

Chapter 6 begins with a discussion of Foucault's history of techniques of power, from the spectacular displays of torture and execution that demonstrated the power of the sovereign in feudal times, to the disciplinary techniques that emerged alongside the rise of capitalism. The argument then builds on the work of a distinctive group of thinkers who have developed a theory of a post-disciplinary society of control, most notably

Deleuze (1995) and Hardt and Negri (2000). The concept of 'biopolitics' – which was originally formulated by Foucault to explain the administrative techniques that emerged during the eighteenth and nineteenth centuries for the control of bodies and entire populations – is developed by these later thinkers to encompass methods developed since the twentieth century for regulating flows of information and the fruits of emotional and intellectual labour. Information societies are moving away from disciplinary methods of social control to non-disciplinary methods of control that focus on the regulation of flows of information, controlling the output of our brains in addition to the output of our bodies. The 'control society' is characterized by techniques for the modulation of flows, following cybernetic principles for control and equilibrium, in contrast to traditional disciplinary techniques that regulate behaviour according to a fixed mould. With the increasing importance of intellectual property and tacit forms of knowledge, new forms of surveillance are emerging to capture and regulate these flows.

Chapter 7 shows how certain representations of time are bound up with different kinds of information technology and methods of social control. Resistance to such control is often characterized by attempts to evade institutional methods of time keeping, such as the timetable and real-time control, and establish a sense of identity independently from these mechanisms. The argument begins by outlining a critique of the quantitative measurable concept of time, which has been used to measure and control the activities of labour since at least the Industrial Revolution. The concept of real-time is the latest development in a series that runs from the clock-time of large-scale manufacturing, to the stopwatch of the time-and-motion study, to the real-time control of computer-mediated work. Real-time is defined as the time of information warfare, where real-time technologies of control are simultaneously transforming both the civilian world and the military field of operations. In contrast to this technologically orientated view of time, a qualitative conception of time will be developed drawing from the critical work of Nietzsche, Bergson and Negri. In common with all these thinkers is an emphasis on time as a fundamentally creative concept, not merely an instrument for measurement and control. Negri's radical re-evaluation of time will be given particular attention. This conception of time recognizes the primary productive forces of cooperation, from which all other forms of organization spring. In the dominant service sector economies, the kind of intense cooperation required and the centrality of communication networks to the productive apparatus makes Negri's conception of cooperative time, or *kairos*, particularly suitable for this analysis. Negri's work is alone in addressing the biopolitical implications of the control society.

There is already much enthusiasm for the topic of information warfare in the United States, particularly in the area of information systems (Denning 1999), and in military academies (Arquilla and Rondfeldt 1997).

There are several overlapping themes concerning the nature of network organizations that are also covered by studies in sociology and business strategy – for instance, Castells' *The Rise of the Network Society* (1996) and Nonaka and Takeuchi's *The Knowledge Creating Company* (1995). *Information Warfare in Business*, is intended to serve as a critical introduction to the practice of information warfare, with particular reference to case studies from business organizations, and the surprisingly creative forms of organization – such as Napster, Linux and the Zapatistas – that have emerged to challenge existing business practices. Many of these organizations have caused businesses to re-evaluate their current business practices, both in terms of the way in which they work and their relationships with their customers and other stakeholders. What is actually at stake with the rise of information warfare? For one thing, one can be the target of information warfare without ever knowing that it has happened. Its techniques are aimed at cherished sources of information and the benefits that derive from access to these sources. Information warfare can take many forms; it can involve identity theft to gain illegal access to personal information and wealth, or it can even be employed to manipulate what we know about ourselves and what we know about the world around us. Its methods threaten our ability to respond to changes in the environment and our very identity. This is the case for business organizations, no less than for individual citizens or nation states.

Each of these chapters approaches the practice of information warfare by identifying points of resistance, where a struggle is defined largely in terms of the protection or targeting of information resources. Such struggles have emerged over recent years surrounding our access to information, the ownership of knowledge, surveillance at work, the regulation of intellectual property and the security of communication networks. Our sources of information and communication networks can be attacked by various means, from low-tech approaches such as rumour and gossip, to more high-tech approaches such as the computer virus. We can not tell in advance how things will turn out, or what new forms of productivity and creativity will emerge from our evolving networks of cooperation and organization. Neither can we tell how techniques of control will develop in the future to capture and regulate these creative efforts. Deleuze conceived of the future as a line of flight (Deleuze 1995, Deleuze and Guattari 1988). This flight emerges out of existing struggles, but in a movement that undoes established territories, often making a weapon of necessity. The Zapatistas developed techniques of information warfare to counter the superior power of the Mexican state. Linus Torvalds dreamt up Linux as a Masters student in computer science due to his concerns about the prohibitive expense of existing computer operating systems. The distributed infrastructure of the Internet was not directly the consequence of a military research programme to defend against nuclear attack; indeed, the Pentagon rejected this proposal as a plan for the ARPANET at an early stage in its development. Although the architecture of the ARPANET originally seems to have been the optimization of costly online time sharing between scientists, its distributed architecture emerged from scientists who used the network for the exchange of electronic mail, which was especially popular with fans of science fiction (Castells 2001). The line of flight is not the same as a 'research programme'; it is an uncontrolled experiment that tends to upset the existing status quo, forcing us to re-evaluate our cognitive maps. Only afterwards can corporations and the state domesticate and capitalize on the innovations produced by different lines of flight. Information warfare is such a line.

# Part I

# Creative organizational forms

# 2 Strategies for the Information Age

## Nomadic wisdom and guerrilla warfare

> They were as unstable as water, and like water would perhaps finally prevail.
>
> (T.E. Lawrence, *The Seven Pillars of Wisdom*)

This chapter examines the innovations in organizational strategy that have been emerging over recent years that take advantage of the new communications technologies. In the early 1990s, management gurus such as Michael Hammer spoke of how information technology could be used to 're-engineer' the corporation, not just in terms of information processing but by completely redesigning the organization around its IT. A few years later the sociologist, Manuel Castells (1996), described the emergence of the 'network society', which was a global society where many of the traditional hierarchical forms of organization were breaking down. According to Castells, organizations were increasingly coming to resemble networks rather than hierarchies. In the first volume of his trilogy on the network society, Castells chose Cisco Systems as his *example par excellence* of the network enterprise, a firm that had a massive intranet of its own and was the world's largest single producer of networking technology. In the second volume of this trilogy, which was devoted to the wider social changes resulting from the growth of communications networks, he focused not on a high-tech organization from Silicon Valley, but on a remarkable band of guerrilla fighters from an impoverished region of Mexico (Castells 1997). He did this because these rebels had cleverly exploited the Internet and other communications media to turn their rebellion into a war of words fought in the international arena, rather than what might have ended up as a short-lived local peasant revolt. This chapter builds on Castells' initial insight by showing that many organizations are emerging that take as their model the guerrilla fighter rather than the more territorial market-orientated approaches. These new strategies have much in common with what the RAND Corporation are now calling 'information warfare' (Arquilla and Ronfeldt 1997). Nomadic and guerrilla strategies not only explain the ongoing revolution in military thinking, but are equally suitable to describe many other kinds of organization.

The discussion begins by highlighting the territorial assumptions that underpin much of the thinking in strategic management. In contrast to this, a nomadic approach to strategy is outlined that does not assume the acquisition of property or new territories as its goal, but which can still sustain a community and a specific way of life. The idea of a nomadic strategy will be introduced with a discussion of the history of ancient nomadic tribes, and then elaborated with reference to the work of T.E. Lawrence. Lawrence's writings on the subject of guerrilla warfare provide many useful insights into how he combined the nomadic lifestyle of the Bedouin with guerrilla tactics to great effect in his desert campaigns during the First World War. These insights will be distilled in terms of a number of principles that can be summarized under the following names: leadership, friendship, the idea, peace and non-conflict, the pack, calculation, speed and communication. The main thrust of this chapter reveals that these principles can be seen in all sorts of new social organizations emerging today, specifically in relation to the new communications technologies. A number of cases will be discussed, each of which powerfully illustrates the effectiveness of nomadic wisdom and guerrilla strategies, including groups as diverse as the Zapatista rebels in Mexico, the Linux and Napster Internet communities, and the Slow Food Society. This argument draws on a variety of literatures, spanning fields including ancient history, military strategy, business strategy, sociology and contemporary developments in the modern information technology.

## Strategy and invasion

Mainstream approaches to strategic management take the pursuit of wealth as a given and so focus on developing an organization's competitive advantage, as illustrated in the work of best-selling authors such as Michael Porter (1980, 1985, 1990). This literature on strategic management has not paid much attention to strategies that take a 'nomadic' orientation, and any such discussion of nomadic kinds of strategy has been marginal at best. This is primarily because a key principle of nomadic strategies is that the acquisition of wealth or property is not their primary aim, and in fact, they may even be opposed to acquisition of this kind. Competitiveness does not really apply to nomadic strategies, since it assumes that people are competing in the same game, each wanting to protect and develop their respective territories, but, as we shall see below, things are often not so straightforward.

The literature on strategic management is littered with military metaphors and analogies, especially with regard to defending your native territory and setting out to conquer new territories. Cummings (1993) has observed that the term 'strategy' has military origins and derives from the ancient Greek 'strategos', who were the military defenders of the city of Athens. Cummings states that the successful business leader possesses the same virtues as the successful military leader. The business leader is just as important as the leader in battle – both are supposed to inspire their own men and out-manoeuvre and defeat their opponents. More mathematical approaches to strategy, such as game theory, have been taken wholesale from military strategy, where a direct parallel is made between the 'game' of war and that of business (Ho and Weigelt 1997). Michael Porter's work uses exactly the same kind of analogies; for instance, he advises managers that, 'A maxim of defense in military strategy is that it is extremely costly to defend an entire perimeter against attack from any direction where the challenger may employ any weapon. The same principle applies in competitive strategy' (Porter 1985: 504). Elsewhere he writes that, 'defense can be achieved by forcing competitors to back down after a battle. However, the most effective defense is to prevent the battle all together' (Porter 1980: 98). His idea of competition is a military one, and he explicitly refers to 'competitive warfare' (Porter 1980: 89, 274). He describes his competitive analysis in terms of 'offensive' and 'defensive' moves, on 'picking the battleground', and of methods of 'deterrence' and 'retaliation' (Porter 1980, 1985). In this analysis, the market segment is the territory over which war is waged. Indeed, a minimal definition of the concept of territory is 'any defended area' (Bogue 1999), which is precisely how Porter describes a market segment. The five forces by which a territory is marked out includes the bargaining power of suppliers, the bargaining power of buyers, the threat of substitutes, the erection of barriers to entry, and the rivalry between the existing competitors. By looking at the composition of the five forces at work in his or her particular industry, a manager should be able to protect his or her own territory, by means such as setting up barriers to entry or by keeping buyers loyal rather than switching to a substitute or competitor. Porter (1990) recommends a number of strategies to deal with one's rivals or opponents, such as making new acquisitions in order to enter new markets and the formation of alliances to protect or extend one's territory. This latter approach is seen as a last ditch weapon, because we are warned that alliances can weaken our desire to compete. Strategy is thus a territorial affair, where the principal aim of a strategy is the defence of existing borders and the expansion into neighbouring lands.

Porter (2001) has investigated the implications of the Internet on corporate strategy, in an exceptional article that he published in the *Harvard Business Review*. In this article, he took pains to criticize the existing hyperbole surrounding the dramatic rise in interest in the so-called 'dot.com' companies and the inflated prices of high technology stocks. He criticized the lexicon of jargon that emerged to explain the apparent success of the dot.com companies, such as 'e-business', 'e-strategy', 'business model' and 'new economy'. This was accompanied by a proliferation of creative accounting techniques, which downplayed traditional measures of value, such as profitability, and incorporated new measures, such as number of customers, number of website visitors and so on. Porter argued that there was no 'new economy'. As things turned out, the high technology market collapsed, and we know that his critique was spot on. According to Porter, the Internet has had a minimal impact on corporate strategy, and there is no reason to abandon the proven principles, which have already been outlined in some detail in his previous works. He does concede to some minor innovations, none of which has had a major impact on strategy as such. One is the creation of the new industries, for example the online auction and digital market places of eBay and Amazon, that open onto a huge potential market. Another involves using real-time technologies to re-engineer the entire value chain, from the supplier all the way to the customer. However, this should not be seen as source of competitive advantage since the technology is widely available, making it impossible to sustain any temporary advantage. Investors had clearly misread the signs; in Porter's words, 'Rather than signalling a healthy business environment, the sheer number of dot.coms in many industries often revealed nothing more than the existence of low barriers to entry, always a sign of danger' (Porter 2001: 65). This statement reveals much about the Porter's territorial analysis of industry. Again, the market is described in terms of a defended area, from which outsiders must be excluded, 'the ability to create barriers to entry is critical' (Porter 2001: 70). Only once such barriers have been erected can the managers 'capture the economic benefits the Internet creates' (Porter 2001: 64). One need not dispute the intelligence of Porter's analysis to highlight its territorial suppositions and its acquisitive emphasis. However, alternative approaches to strategy exist that are almost entirely neglected in Porter's work, and which are not grounded in the same apparatus of capture. These alternative approaches to strategy may be compared with the nomad warrior or guerrilla fighter, based on free movement rather than on capture and regulation. The kinds of organizations that have adopted such strategies are by no means a dominant force within business, but they are nonetheless important exemplars of a distinctive network form of organization. Furthermore, these nomadic networks are the very same organizations that have been pioneers in the practice of information warfare. Before proceeding with a detailed analysis of the importance of nomadic strategy, more needs to be said about the concept of territory in business strategy.

Territorial thinking is by no means restricted to mainstream rhetoric, but is also to be found in more critically inclined academic work. Perhaps the most sophisticated expression of this territorial kind of thinking in the management literature is the work of Karl Weick, who has been a chief proponent of the use of maps for thinking, specifically for strategy formulation (Weick 1994, Colville *et al.* 1999). Weick suggests that maps can perform several functions; they can provide a sense of security in uncertain times, and they can motivate action. According to Weick, the accuracy of a map is far less important than the mere fact of having one, because either way it stimulates adaptive action. To illustrate his point, Weick recounts a tale from the poet Miroslav Holub about a group of Hungarian soldiers who had become lost in the Alps. In this story, the soldiers had become disorientated and had given up hope of rescue when one of the party found a map in their pocket. This got them moving again and eventually they found their way back to their camp. However, on arrival they discovered that the map they had been using as their guide was not a map of the Alps at all, but a map of the Pyrenees. Weick draws a number of lessons from this story about the socially constructed nature of the human world. In some sections, Weick says that maps are entirely socially constructed and experimental, whereas in others he states that maps can be more or less accurate and as such presumes that they are models of the world (even if this is also socially constructed). But this leads to some serious contradictions where, on the one hand, he says that in a socially constructed world the map creates the territory (Weick 1994: 213), and yet on the other hand that maps can be more or less accurate approximations of the territory (Weick 1994: 213, 219). It is clearly nonsense to say that the map creates the territory and then that it is a more or less accurate representation of it (i.e. itself). A generous interpretation of this contradiction might be that he is following some kind of dialectical procedure. If this is the case, it is difficult to determine the movement from one term in the dialectic (the map approximates the territory) to the other (the map creates the territory), or vice versa. At the very least, some clear relation must be drawn between these very different terms, that of representation and that of creation.

In fact, Weick's map is less experimental and exploratory than it is imperialist and invasive, and he says as much himself. He explicitly draws a comparison between managing and the US invasion of Grenada stating that, 'Managers invade new markets before the cartographers hand them a map, and before people are sure the invasion is legitimate. The invasion becomes the pretext to learn what is being invaded and what constitutes legitimate grounds for the invasion' (Weick 1994: 215). It seems odd to question what the legitimate grounds for invasion might be after an invasion, although possibly in keeping with US foreign policy. Chomsky (1992) asks, for example, why it was legitimate for the USA to invade Grenada and countless developing countries, but it was illegitimate for Iraq to invade Kuwait? One suspects that the idea of a legitimate invasion probably depends on whose side you happen to be on. In this respect, Weick shows a clear affinity for those who hold positions of power within organizations, and the mapmakers to whom he refers are almost exclusively managers. Weick states that the map 'creates the territory, labels the territory, prefigures self-confirming perception and action' (Weick 1994: 213).

This activity of invading, labelling and thereby creating territories is the real job of management. Labelling things and territories has always been province of the powerful. Nietzsche's genealogical approach to language makes this quite clear to us:

> The seigneurial privilege of giving names even allows us to conceive of the origin of language itself as a manifestation of the power of the rulers: they say 'this is so and so', they set their seal on everything and every occurrence with a sound [or sign] and thereby take possession of it, as it were.
>
> (Nietzsche 1994: 13)

There is a curious ambiguity in Weick's work concerning the extent to which people are either creative or docile. In some places Weick describes managers as the invaders and visionaries of the company, but in others he portrays them as fairly docile and in need of reassurance, adaptive rather than creative. According to Weick, the active element of mapping emerges when it provokes people into action and stimulates movement; this seems to suggest a basic appreciation of nomadic behaviour. However, he also writes of mapping as a way of controlling people's movements. In his own words, 'people need abstractions to smooth over the differences. People need to be cartographers in order to fashion those disconnected abstractions into more plausible patterns . . . people need to adopt the myth that their maps are a sufficiently credible version of the territory. . .' (Weick 1994: 218–219). Weick resorts to a functionalist and very questionable description of people here, detailing what he believes people need over and over again. People need the reassurance of the familiar (patterns), they need to smooth over the alien and unfamiliar (differences), they are purely adaptive rather than creative beings. In a subsequent article, even more emphasis is placed on the pacifying effects of maps (Colville *et al.* 1999). Weick, therefore, is less concerned with the stimulation of free nomadic movement than he is with the regulation of movement which 'imposes order on the situation' (Weick 1994: 219). The kind of activity encouraged by using the map might be better termed as reactivity, because the map 'stimulates the adaptive process' – not the creative, or experimental process, to say nothing of critique.

In contrast to the kind of acquisitive and territorial approaches to organization discussed above, the rest of the chapter will be devoted to outlining a nomadic approach to strategy and organization. One of the general principles of nomadic strategies is that they do not take the acquisition of wealth or property as their main aim. In fact, some nomadic forms of organization may even be opposed to acquisition of this kind. Competitiveness does not really apply to nomadic strategies, since it assumes that people are competing in the same game, each wanting to protect and develop their respective territories. A good example of this nomadic approach to strategy is the Linux community, because Linux does not compete with Microsoft in the traditional sense. The main goal of Linux is not directed at the accumulation of profits but concerns the ideals behind Linux itself, such as the development of open source programming and the dissemination of free software (Moody 2001). The Linux community is not an organization comparable to a traditional business corporation and is not bound by the same rules. Instead, it is a dispersed community of volunteers, who are trying to change the rules of the game rather than trying to compete on the terms dictated by Microsoft. Simple competition is therefore an inadequate term to explain this kind of strategy. Before discussing this case in any detail, the next two sections will outline a general theory of this kind of nomadic strategy first by looking at the wisdom of the ancient nomadic tribes of the Scythians, and then by examining the principles of guerrilla warfare developed by T.E. Lawrence.

## Nomadic wisdom: strategy as movement

Weick describes strategy as a matter of mapping movements and regulating people's movements. He describes strategy in terms of planning invasions and the need for security. However, this somewhat imperialistic understanding of strategy can be contrasted with a nomadic kind of strategy that explores the power of unregulated movement. This nomadic style of strategy has been summarized in John Sellars' (1998) study of the ancient Scythians and is discussed in more detail in T.E. Lawrence's (1962) compelling account of the Arab Revolt during the First World War.

The Scythians were a group of loosely organized tribes that roamed the southern Russian steppes during antiquity. Sellars (1998) begins his account of these tribes by looking at the difference between city dwellers (*polis*) and the wandering tribes (*nomos*). The Greek term *nomos* originally designated the steppe or pasture, but later came to refer to customs and later still to the law. However, it is with reference to the pasture or steppe that the term nomad was derived from *nomos*. The historian Herodotus praised the Scythians for their wisdom, and their skills in self-preservation. This wisdom consisted partly in the fact that they carried few possessions and were an unattractive proposition to potential invaders, but also they were always on the move and so had little to defend. As Rosen explains, 'they [Scythians] cannot be captured in war, while their enemies necessarily perish in the act of pursuing them' (Rosen 1993: 36). Their strategy was, in fact, their way of life. This strategy was so successful that at one time or another they conquered large parts of Europe and Asia, and they endured on the steppes from 900 BC to 200 BC, a period longer even than the life of the Roman Empire.

The city dweller and the nomad each have their own specific mode of distribution, which underpins their distinctive way of life. In the city state, land is parcelled up and distributed to the people, whereas, for the nomads, people and animals distribute themselves across the land as they move. The nomadic tribes were chiefly animal-raising peoples and Sellars (1998) suggests that this is fundamental in understanding their movement as hordes or packs. He contrasts these two ways of life as follows, 'Whereas the State requires an extra dimension, giving it an external point of view from which it can enforce its sovereignty, the nomadic pack operates immanently and directionally, moving in the manner of a vector, following a flow' (Sellars: 74–75). Looking back at the previous section on Weick, we can see that his work is clearly opposed to nomadic thinking. For Weick, a map generates an external point of view from which management can exert its sovereignty.

Identifying the nomads can be a singularly difficult thing to do, and for good reason. One might think of Nietzsche's observation that the very act of naming (the 'seigneurial privilege' as he termed it) is a means of taking possession of a thing, and it is precisely this kind of enslavement that the nomad wishes to avoid. Sellars' essay admits that, 'there are a number of difficulties in trying to define the Scythians. They were never a nation nor a race, but rather a multiplicity of tribes sharing a similar way of life' (Sellars 1998: 86). Minns study of the Scythians observed that the term 'Scythian' has no ethnological meaning, but is primarily a *political* term referring to a particular nomadic way of life (Minns 1913: 35). Unlike settled tribes, nomadic peoples tend not have written histories and are defined more by their geography rather than by their history (Deleuze and Guattari 1988). It is clear that movement is fundamental to the essence of nomadic life but not just any kind of movement. Nomads are often reluctant to move outside their territory and have a unique way of occupying it. Unlike the highly regulated movements of 'civilized' city dwellers, nomadic movement is unregulated. Sedentary space is divided up, regulated and distributed to the people, but nomadic space is traversed by the people in the same manner as a pack of animals. In fact, it is the search for new pastures to graze their animals, which forces the nomads to move alongside their herd. The regulated movements of city folk appear to originate as if from a force that is external to them, whereas nomads are driven as if from an inner motor. Deleuze and Guattari have suggested that the nomad is defined more by the idea of speed than simply in terms of movement. The nomads occupy their territory by means of their ability to pick up their things and leave at great speed. Whereas movement is defined in a purely extensive manner, the concept of speed has an intensive dimension. The intensity of speed is characterized not in terms of a measurable distance travelled in a given period, but by its qualitative features. For example the nomadic quality that Lawrence borrowed from the Bedouin was the ability to spring up anywhere. The speed of travel of the nomad on horseback can also be broken down into other distinctive qualities, such as the trot, cantor or gallop. Speed may be slow or fast, but has an inner intensity, like that of the martial arts which are as much about suspense and immobility as they are about the speed of attack. There are times when the pack may pause at a watering hole or wait for a pass to clear, but at other times they move rapidly, fleeing danger or attacking without warning. It should be noted that a nomadic strategy is not necessarily suitable only for nomadic peoples, but is a way of thinking about strategy in terms of speed and movement. Thinking itself can be a way of moving.

In sum, the nomad's way of life is their strategy. This is a life, or strategy, defined by movement, where the tribe travels as a pack and holds few possessions by means of which it could become enslaved. It defends its territory not by means of settlement and fortification but through movement and speed. This kind of nomadic strategy is by no means restricted to ancient nomadic peoples and T.E. Lawrence developed a similar strategy as part of the Arab Revolt against Turkish occupation in World War One.

## Principles of a nomadic strategy: the Arab revolt

> We reminded one another that movement was the law of strategy, and started moving
>
> (Lawrence 1962: 489)

Many theoreticians and practitioners have written rich accounts of the principles underlying guerrilla warfare, such as Walter Laqueur's (1977) comprehensive history of guerrilla warfare and revolutionaries, from Che Guevara to Subcomandante Marcos. Lawrence's account has been chosen for a number of reasons, (i) his work provides an extensive study of both the theory and practice of guerrilla warfare, including much rich ethnographic material, (ii) his work was an inspiration to the philosopher Gilles Deleuze whose concepts of the rhizome and smooth space in turn provide the conceptual grounds for the present chapter, and (iii) his work emphasizes the role of ideas and communications systems in guerrilla strategy, which are key features of the case studies discussed later in this chapter. Lawrence himself observed that advances in communications technology were continually changing the conduct of warfare and rebellion: 'The printing press, and each newly-discovered method of communication favoured the intellectual above the physical, civilization paying the mind always away from the body's funds' (Lawrence 1962: 201).

T.E. Lawrence was a map maker for the British Army before he became embroiled in the Arab Revolt. By working with the nomads, he invented a hybrid war that transformed the nomadic lifestyle of the Bedouin into an unusual form of warfare. He frequently criticized the army's tactics, which relied heavily upon drawing a map of the territory and then occupying strategic positions within that territory, calculating resources needed, acceptable losses, and so on. Lawrence went about problem solving differently; he accepted no losses as necessary, resolved to fight no battles, and hold no territory. Instead he pioneered a form of desert guerrilla warfare. Lawrence outlined a number of principles for nomadic warfare in his extensive and brilliant book, *The Seven Pillars of Wisdom* (Lawrence 1962). The following list has been derived from Lawrence's description of the Arab revolt.

- *The Principle of Leadership*
  The presumption that a leader could easily direct the revolt was an utterly foreign and inappropriate concept. Lawrence himself stated that he was caught up in events and only rarely could he have things as he wanted them: 'My proper share was a minor one, but because of a fluent pen, a free speech, and a certain adroitness of brain, I took upon myself, as I describe it, a mock primacy' (Lawrence 1962: 22). The rebellion was a function of the nomadic way of life where lines could be broken at any point, there was no control centre: '...we might be a vapour, blowing where we listed...' (Lawrence 1962: 198).
- *The Principle of Friendship*
  This principle is closely allied with that of leadership, in the sense that friendship may be a far more important driver of rebellion than the power of a leader. In this light, there could be no idea of acceptable casualties, 'an unnecessary action, or shot, or casualty, was not only waste but sin. Our rebels were not materials, like soldiers, but friends of ours...' (Lawrence 1962: 168). Guerrilla warfare is a people's war.
- *The Principle of Speed*
  Rebellion involves constant movement, speed, and the ability to spring up anywhere: 'Our cards were speed and time, not hitting power' (Lawrence 1962: 201–202). Lawrence discovered that the power of nomadic strategy lay both in the capacity to lie low and wait for long periods of time, but also in the ability to attack and retreat at great speed when necessary.
- *The Principle of the Idea*
  The role of propaganda and intelligence was frequently emphasized by Lawrence. Movement is therefore crucial both on a physical and an ideological level, where rebellion was embodied as an idea: 'Our kingdom lay in each man's mind...' (Lawrence 1962: 198). In the final analysis, the primary object of guerrilla war was the territory of the mind, 'A province would be won when we had taught the civilians in it to die for our ideal of freedom' (Lawrence 1962: 202). Lawrence also observed that it may be more difficult for an enemy to destroy an abstract idea than to destroy a physical body, 'suppose we were ... an influence, an idea, a thing intangible, invulnerable, without front or back, drifting about like a gas?' (Lawrence 1962: 198). Despite being a key strategic principle, this was the source of a great deal of anguish for Lawrence due to his ambiguous role in the rebellion. He explained his difficulty in following two masters as follows, 'I risked the fraud, on my conviction that Arab help was necessary to our cheap and speedy

victory in the East, and that better we win and break our word than lose' (Lawrence 1962: 24). As a consequence Lawrence was, '. . . continually and bitterly ashamed' (Lawrence 1962: 77).

- *The Principle of Peace and Non-Conflict*
  Rebellion is a form of peace rather than a form of war (Lawrence 1962: 149). There were few battles in the revolt and no fixed points of conflict: 'We had nothing material to lose, so our best line was to defend nothing and to shoot nothing' (Lawrence 1962: 201). Lawrence wrote that his kind of guerrilla rebellion had more in common with a national strike, rather than overt physical combat. Guerrilla warfare relies greatly on the assistance of the civilian population and has as much to do with passive resistance as it does with direct conflict. Lawrence believed this to be a major point of departure between his own approach and the most influential theories of the day, such as those of Clausewitz and Foch. This is also closely allied with the principle of friendship stated above.
- *The Principle of the Pack*
  Since the revolt was not the work of a disciplined corps of well-equipped fighting men, there was little point in acting as if it were. Instead, raiding parties were brought together as small packs of fighters on the principle that: 'The smaller the unit the better its performance' (Lawrence 1962: 140). Lawrence also drew a comparison between the desert and the sea, where camel raiding parties were analogous to ships that could cruise the enemy's frontiers and then disappear without trace when danger threatened (Lawrence 1962: 345).
- *The Principle of Calculation*
  Calculations relating to a disjointed irregular army were far more complex than those for a disciplined mass of troops. The power of a traditional army may be calculated to a large extent as comprising homogeneous units, in terms of the number of men and armaments. In contrast, the power of guerrilla fighting units must be determined in conjunction with their geography and nomadic lifestyle. As Lawrence explained, 'The efficiency of our forces was the personal efficiency of the single man. It seemed to me that, on our articulated war, the sum yielded by single men would at least equal the product of a compound system of the same strength' (Lawrence 1962: 348). This might also be called the principle of the assemblage, because it highlights the potential of combinations of qualitatively different factors: man-camel-desert-rifle, ideas-printing press-territory. Considered as an assemblage, the freedom of movement of a guerrilla fighter or an idea is impossible to capture in terms of simple abstract calculations.
- *The Principle of Communication*
  The rebels made it impossible to regulate communication flows within the territory by cutting off lines of communication and creating rapid,

> temporary lines. The desert became a space that could never be striated or regulated. The ability to disrupt the lines of communication was far more important than any numerical advantage: 'The death of a Turkish bridge or rail, machine or gun or charge of high explosive was more profitable to us than the death of a Turk' (Lawrence 1962: 199).

T.E. Lawrence's approach to strategy was to make it impossible for others to hold a territory rather than claim any for himself. He did not try to represent the Arab tribes but acted as a relay for them, a go-between for the different tribes and the British. The principles for revolt that were set out by Lawrence share many of the same characteristics as the rhizome described by Deleuze and Guattari (1988). In contrast to the tree, which exemplifies rigid hierarchical organizations, the rhizome is a kind of bulb or tuber that is composed of intensive states and has little or no internal structure. Any point in a rhizome can be connected to any other. It can be broken at any point, but can re-grow along one of its older lines or set out in a new direction entirely. It is characterized by lines of flight rather than fixed points through which one must pass. It has no underlying model or pattern, and develops through experimentation with what it comes into contact. The rhizome is a multiplicity, an assemblage of qualitatively different elements that combine to give it freedom of movement. In Lawrence's terms, his fighting force was an assemblage composed of 'the Algebraical element of things, a Biological element of lives, and the Psychological element of ideas' (Lawrence 1962: 197). Only when considered in connection with a specific geography, climate, culture, weaponry, means of transportation and communication, could the full striking power of a man or pack of men be understood properly. This assemblage of qualitatively different factors gave the irregular forces their ability to occupy space, and eventually allowed them to defeat an adversary that was numerically far superior. The rhizome metaphor thus provides a very good way of understanding the organizational approach of guerrilla warfare outlined above, in contrast to the tree-like structure of the regular army.

## Nomads and guerrillas: a way of life or a means to an end?

> ...walking is a form of fighting and to avoid combat at a given moment is also another form.
>
> (Guevara 1998: 154–155)

The work of T.E. Lawrence places particular emphasis on the suitability of nomadic life for waging guerrilla warfare, and one might suggest that the nomad's strategy is a reflection of this way of life. As a way of life, nomadic strategy should not be seen as requiring violent conflict. It only becomes violent when it meets a wall, an opposing force that attempts to fence it in. In fact, it could be argued that nomadism is more concerned with innovation and art, in contrast to the nation state, which uses violence to maintain the integrity of its borders and regulate movements within them (Deleuze and Guattari 1988). Nomadic strategy is not identical to guerrilla strategy, although there are many remarkable coincidences. For example, almost every principle outlined above can also be found in a slightly different form in Che Guevara's (1998) work on guerrilla warfare. However, this work is quite different from Lawrence's because Guevara saw nomadic strategy more as a means to an end rather than as a way of life in its own right. Guevara's work describes guerrilla warfare as a 'nomadic stage' after which the state reasserts itself under the guise of communism (Guevara 1998: 34, 44, 62). He explained the temporary nature of this strategy in the following terms: 'guerrilla warfare is a phase that does not afford in itself opportunities to arrive at complete victory.... Triumph will always be the product of a regular army, even though its origins are in a guerrilla army' (Guevara 1998: 13). Guerrilla warfare may borrow much from nomadism, but it is not identical to it.

The most important concepts that guerrilla warfare borrows from nomadism are the priority of speed and the creation of a space within which the nomad/guerrilla can move unimpeded. Guevara observed that this kind of movement can be facilitated by using nightfall as a way of concealing one's whereabouts (Guevara 1998: 19), and by operating 'in places beyond the reach of repressive forces', such as in dense forests, steep mountains, impassable deserts or marshes (Guevara 1998: 9, 25). He also emphasized the importance of secrecy and invoked the metaphor of the rebellion as an 'underground movement'. This is a remarkably apposite metaphor because the 'underground' can be defined precisely as a territory that exits underneath and invisible to the surface territory, which is controlled by the forces of the state. Almost every principle of nomadic strategy that was outlined by Lawrence can be found in some form in Guevara's own manual of guerrilla warfare. This can be seen clearly from the following list, where the eight principles of nomadic strategy that have been derived from Lawrence's work earlier in this chapter will be compared with Guevara's own guerrilla theory.

- *The principle of Leadership*
  Guevara tended to focus more on formal leadership roles than Lawrence. However, he still distinguished the hierarchy of his guerrilla forces from a traditional hierarchy of 'tyranny' (Guevara 1998: 55). He also noted that the form of guerrilla organization is not driven by traditional social structures but by the specific cultural and geographical conditions: 'No rigid scheme can be offered for the organization of a guerrilla band...' (Guevara 1998: 54).
- *The Principle of Friendship*
  Like Lawrence, Guevara emphasized the importance of informal group relations in guerrilla warfare, where, 'The guerrilla soldier must

be an extraordinary companion' (Guevara 1998: 43). Friendship also played an important role in maintaining a relationship between guerrillas and the local community, '...he [the guerrilla] should preferably be an inhabitant of the zone. If this is the case, he will have friends who will help him...' (Guevara 1998: 41).

- *The Principle of Speed*
  Guevara stressed the importance of speed on numerous occasions in his account of guerrilla tactics as it gives the guerrilla band one of their few advantages over the regular army, '...it is always possible to carry out guerrilla attacks in such as way as to assure surprise; and it is the duty of the guerrilla fighter to do so. "Hit and run" some call this scornfully, and this is accurate' (Guevara 1998: 13). Speed of movement can give an enemy the impression that they are encircled, even by numerically inferior forces, '...the fight is started at any one of the points, and the army moves toward it; the guerrilla band then retreats, always maintaining visual contact, and initiates its attack from another point' (Guevara 1998: 19). Speed also opens up greater opportunities for escape after an encounter with the enemy '...the guerrilla fighter needs a good knowledge of the surrounding countryside, the paths of entry and escape, the possibility of speedy maneuver, good hiding places...' (Guevara 1998: 10–11).
- *The Principle of the Idea*
  Propaganda plays an important role in the guerrilla strategy where the rebel '...fights in order to change the social system that keeps all his unarmed brothers in ignominy and misery' (Guevara 1998: 10). Guevara recommends that rebel soldiers carry inspiring or useful books on their travels, which describe past heroes or local histories (Guevara 1998: 53). This helps keep morale high and Guevara believed that it might also ward off undesirable pastimes such as gambling: 'People with such notable devotion and firmness must have an ideal that sustains them in the adverse conditions that we have described' (Guevara 1998: 46). Guevara also recommended the use of radio broadcasts to spread their alternative message as rapidly and as widely as possible.
- *The Principle of Peace and Non-Conflict*
  Unnecessary violence is condemned for being wasteful and provocative of severe reprisals against innocent people. Non-violent measures, such as sabotage, are to be preferred: 'Acts of sabotage are very important. It is necessary to distinguish between sabotage, a revolutionary and highly effective method of warfare, and terrorism, a measure that is generally ineffective and indiscriminate in its results, since it often makes victims of innocent people...' (Guevara 1998: 21). Like Lawrence, Guevara recommended the general strike as a means of resistance in the later stages of rebellion: '...it is possible to arrive at organized mass action in centres of work, of which the final

result will be the general strike. The strike is a most important factor in a civil war...' (Guevara 1998: 16). Lawrence, however, gave much more weight to the analogy between guerrilla warfare and peaceful means like the general strike. On the whole, Guevara was much less hesitant than Lawrence in prescribing violence, which he saw as necessary to provoke the true nature of the repressive state (Guevara 1998: 147).

- *The Principle of the Pack*
  At the core of Guevara's analysis is the pack: 'The guerrilla band is an armed nucleus, the fighting vanguard of the people' (Guevara 1998: 10). His analysis of guerrilla warfare devotes several pages to the ideal size of the irregular fighting unit.
- *The Principle of Calculation*
  This principle of calculation does not feature much in Guevara's work, although it does emphasize the importance of factors such as equipment, local geography and culture. The closest Guevara comes to recognizing the importance of the assemblage is when discussing the adaptability of the guerrilla: 'With the help of his natural adaptability, he becomes a part of the land itself where he fights' (Guevara 1998: 44), and, 'The facility with which the guerrilla fighter can perform his function and adapt himself to the environment will depend upon his equipment' (Guevara 1998: 49).
- *The Principle of Communication*
  Guevara explained the crucial role of communication in terms of setting up a secretive network of alternative communication channels (Guevara 1998: 16, 28, 34). At the same time the rebel needs to disrupt the lines of communication of the enemy forces (Guevara 1998: 17, 24). The lines of communication are highlighted as being the weakest point of an enemy's infrastructure (Guevara 1998: 23), which can be attacked relatively easily since these cannot be defended along their entire length. This principle is very closely related to two of the principles outlined above, namely the principle of non-conflict and the principle of the idea, all of which derive from the primacy of communication networks to both social life and military organization. In Guevara's own words, 'It is possible to paralyze entire armies, to suspend the industrial life of a zone, leaving the inhabitants of a city without factories, without light, without water, without communications of any kind, without being able to risk travel by highway except at certain hours' (Guevara 1998: 17).

It is clear that Lawrence and Guevara's works share many similarities. They both highlight the crucial role of speed, movement, ideas and communications in their strategies. However, they also differ in significant respects. Despite adhering to very different ideologies (right and left wing) both respect the potential of nomadic strategy, but whereas Lawrence's work emphasizes nomadism as a way of life, Guevara's work views it as a means to an end. Both writers were also steeped in the practicalities of warfare and the principle of avoiding conflict, and both offer very useful guidelines for understanding the potential of innovations in communications networks. Given these general principles, a number of cases will be examined that highlight that the principles of guerrilla strategy are to be found at work in many highly innovative social movements and organizations today. This includes the emergence of online communities such as Linux and Napster, and local networks such as the Zapatista rebels in Mexico and the Slow Food Society in Italy, which are exploiting modern communications networks as part of a general strategy to maintain their indigenous culture and traditions against the influence of powerful multinational corporations.

## The Zapatistas: informational guerrillas

The Zapatistas are a rebel group in Mexico whose membership comes largely from the impoverished indigenous communities of that country. The group has evolved as part of a resistance movement against the kind of economic liberalization that led to the North American Free Trade Agreement in 1994. This liberalization has had a disastrous effect on the fragile economy of the peasant farmers, since it entailed the end of restrictions on corn imports and the elimination of price protection on important cash crops such as coffee (Castells 1997). President Salinas had also reformed article 27 of the Mexican Constitution, ending communal ownership of land by villagers in favour of full commercialization of all land and property. After the peaceful demonstrations by the peasant farmers had been ignored, they began to turn to more violent means, and the Zapatistas were born from this struggle, preparing themselves for a guerrilla war. The distinctive feature of the Zapatista uprising was their use of a range of telecommunications technologies such as video technology and the Internet. Castells (1997: 80) explains their success in the following terms, 'Extensive use of the Internet allowed the Zapatistas to diffuse information and their call throughout the world instantly, and to create a network of support groups which helped to produce an international public opinion movement that made it literally impossible for the Mexican government to use repression on a large scale'. This meant that the rebellion could move away from direct conflict and avoid a bloody war. Castells observed that, through the media, economic pressure may have become an important factor since confidence in the stability of the Mexican economy, and hence President Salinas himself, was shaken as a result of the Zapatistas' actions and their increasing international profile. This event has been taken seriously enough to merit study by experts in Information Warfare at the RAND Corporation (Arquilla and Ronfeldt 1997), and for Castells to call the Zapatistas the 'first informational guerrillas'.

The Zapatistas do have an armed wing and, like many freedom fighters before them, they tread a fine line between being labelled either guerrillas or terrorists. The RAND analysts Arquilla and Ronfeldt do not explicitly condemn the Zapatistas as terrorists, although they place them alongside a list of other 'netwar' activists, which includes: 'terrorists, fundamentalists, ethnonationalists, militant single-issue groups, and criminal organizations...' (Arquilla and Ronfeldt 1997: 373). This is not a favourable comparison to say the least. The RAND analysts observe that Mexico is also fraught with 'Internetted drug cartels' (Arquilla and Ronfeldt 1997: 382). What they do not mention are the massive amounts of information technology and high-tech weaponry that have been imported by the Mexican government from the USA in order to fight these drug barons. The Zapatistas appreciate this only too well because the same technology has been turned against them by the Mexican Government in its efforts to pin down and suppress their rebellion (Marcos 2001). There is a serious human rights issue today due to the fact that the same high-tech weapons and methods of surveillance used to counter illegal traffic in drugs and weapons may also be used to suppress legitimate protest. In the Zapatista manifesto, and by their actions, violence appears to be their weapon of last resort, used only when they are faced with desperate circumstances. They have even received the respect of the Catholic Church in Latin America, and Bishop Ruiz has commended them for their preference for peaceful talks rather than violent confrontation. On the other hand, they have been criticized for refusing to disarm and disband when a compromise was offered by the Mexican state. Why, it was asked would they not return to their towns and villages and settle down when offered an olive branch? Their spokesman, Subcomandate Marcos, had a very clear and principled response:

> They offered us many things – money, projects, aid, and when we rejected them, they became angry and threatened us. That is how we came to understand that, by refusing to accept government aid, by resisting, we made the powerful angry; and there is nothing a Zapatista fighter likes more than making those in power angry. So with singular joy we dedicated ourselves to resisting, to saying no, to transforming our poverty into a weapon – the weapon of resistance.
>
> (Marcos 2001: 167–168)

This ideal of poverty is a distinctive part of the ascetic discipline required by nomadic strategy and may allow nomads to move free from the usual attachment to everyday comforts. Nevertheless, this important principle is often overlooked by many analysts of these emerging organizations, for example Arquilla and Ronfeldt (1997). It is an ideal that is at odds in a world where people have become increasingly defined in terms of their wealth and their possessions, where accumulation is the very basis of the economy. It is also an ideal that crops up in diverse organizations, such as Napster Linux and the Zapatistas, allowing as it does a freedom of movement peculiar to the nomad.[1]

One of the main strengths of the Zapatistas is the widespread popular support, which allows them to melt away into the towns and villages if the fighting becomes too intense in the mountains. The support of the people has perhaps been the main driving force and a critical factor of success with respect to their strategy of peaceful demonstration and their push for democratic reforms. The Zapatistas' spokesman, Subcomandante Marcos, also expresses much gratitude for the support from the non-governmental organizations and protestors in other countries who have come together to form a substantial force acting in the name of the oppressed. He explains the structure of this movement in the following terms: 'This intercontinental network of resistance is not an organizing structure; it has no central head or decision maker; it has no central command or hierarchies. We are the network; all of us who resist' (Marcos 2001: 117). The question of leadership in this case is a complex one where there is no organizational structure as such. Marcos himself may have been appointed spokesman for the Zapatistas, and certainly has a great deal of personal influence, however, he in no way claims to direct the movement. The Zapatista network is simply too big and diffuse to manage. Even at a local level the strategic decisions are made by referendum and plebiscite (Marcos 2001: 247). If anything can be said to be leading the Zapatista movement it is the power of the idea of *Zapatismo* itself. *Zapatismo* is both the ideology and the strategy that drives the Zapatista movement, and it has much in common which the principles of nomadic strategy outlined in the preceding section. This strategy will be discussed in more detail in a later chapter on Information Warfare, but the present discussion will round off by pointing out that the exploitation of communication networks has been a pivotal element of the Zapatista strategy. As Marcos (2001: 181) writes: 'We need this network not only as a tool for our social movements but for our lives: this is a project for life, for a humanity that has a right to critical and truthful information'.

## An Information Age strategy: Linux?

In 1988, Linus Torvalds was a computer science student at the University of Helsinki, Finland. At this time, operating systems such as Unix were expensive and could only run on powerful computers, so Linus wanted to develop an alternative that would be free and could be run on smaller computers. He facilitated the development of his project in 1991 by releasing the source code free on the Web. This has allowed the code to be developed by motivated customers and it has opened up a huge source of creativity, that is, anyone on the Internet. There are some potential problems with this approach because, despite the clear advantages of having a vast, unpaid, and yet highly motivated workforce, questions may arise concerning the reliability of hackers and hobbyists and the coordination of updates. Initially, 'Linux' found favour in universities, but with developments in its user friendliness it has also become popular within the computer industry. In 1999, Hewlett Packard announced that it would be releasing applications for Linux, offering worldwide support, which was followed shortly by similar pronouncements from IBM. The fact that it is free and readily available to anyone on the Web has led to an incredibly rapid increase in the number of users from its early days, with millions of users worldwide today. The rapid success of this operating system has been so dramatic that companies such as Red Hat were founded to provide support and services to Linux users, since although it is not possible to charge a fee for the software it is possible to charge for related support services.

Computer software such as Linux is also associated with ideas of 'open source programming' and 'freeware'. Open source programmes originate from a community of developers who leave their source code open so that others can adapt it to their own requirements. The idea behind freeware is a more profound innovation because it represents a revolution in the concept of Intellectual Property. Whereas traditional copyright law allows the owner of the copyright royalties deriving from the use of their idea, freeware is intellectual property that can and must be used free of charge. This kind of intellectual property has been given the name GNU 'copyleft'.[2] Copyleft entailed the invention of a copyright law to accompany free software known as GNU General Public License, first developed by Richard Stallman in 1985 (Moody 2001). This copyright law was designed to ensure that the free GNU software had to remain free even after later versions had been developed. Later programmes must also be passed on free of charge and open to further adaptation, and if it is combined with other normal programmes these also become subject to the GNU GPL copyright (Moody 2001).

Copyleft may become a powerful innovation since it contradicts the prevailing trends, which are transforming knowledge into a commodity. In contrast to this, copyleft helps lower the barriers that surround the free movement of information and allow it to move nomadically once again. There are many other aspects of the development of Linux which share the same principles as those of guerrilla warfare outlined above. The development of the programme is not done by a well-disciplined corps of programmers but by an ad hoc team of hackers and enthusiasts. Not only was Linux developed by distributed networks of enthusiasts, but the code itself was modular in design so that each of its parts could be developed in a piecemeal fashion and then later woven together to form a functional whole. Initially, Linus himself emerged as the natural leader of the pack, where programmers would pass on their own patches for approval, but eventually as the programme evolved Linus began to delegate this work to trusted lieutenants (Moody 2001: 81). The nature of the GNU General Public Licence has meant that, in theory, anyone can develop their own version of Linux: however, in practice, Linus's name has effectively served as the seal of approval for official versions of Linux. The idea of Linux is itself more powerful than any outright confrontation with possible rivals such as Microsoft. Its availability on the Web combined with the fact that it is free has greatly accelerated both its development and its speed of dissemination. In 2001 the Linux website reported over 40 million users since it was first made available on the Net (www.Linux.com). It is interesting to note that the creator of the copyleft idea, Richard Stallman, was dedicated to the idea of freeware (free software) to such an extent that he left his job at MIT in case the administration attempted to interfere with his plans for the GNU licence. His dedication is clearly expressed in the following statement, 'I don't want a house, I don't want to spend a lot of money. If you spend a lot of money then you're a slave of having to make money. The money then jerks you around, controls your life' (quoted in Moody 2001: 28). Linus Torvalds has also highlighted the fact that a healthy disregard for money seems to be a characteristic of the serious hacker. Torvalds has observed that hackers can happily get by on 'Twinkies and Jolt Cola' so long as they feel a part of the hacker community and share their love for computing (Himanen 2001). One might compare such statements with the nomadic wisdom of the Scythians who 'remain invincible and unconquered by outsiders, because they have nothing to be enslaved for' (Strabo quoted in Sellars 1998: 77).

In sum, the strategy developed by the Linux community can be seen as clearly embodying many of the principles of nomadic strategy that were outlined above. The principle of leadership operates because Linus Torvalds is not a leader in any traditional sense. Although he has a say on the authorized releases of Linux, he works alongside friends and colleagues rather than over subordinates. The principle of the pack is embodied in the loosely knit cellular structure that characterizes the community of programmers who have developed Linux. The principles of speed and communication are apparent in the rapid development of both the workforce of dedicated programmers across many countries and the rapid distribution of the product to the final users of the software. A recent article in the magazine *BusinessWeek* accurately described Linux in the following terms, 'The Linux phenomenon spreads like water – finding its way into all sorts of nooks and crannies. And that's by design' (*BusinessWeek* 3 March 2003). This echoes the sentiments of the quote from T.E. Lawrence that heads this chapter, describing his guerrilla campaigns as 'unstable as water' and powerful for this very reason. The principle of non-conflict becomes apparent from the fact that Linux does not play the same game as Microsoft, i.e. profit, and if it can be said to compete it does so on its own terms. The principle of the idea operates very powerfully, not only with respect to the power of Linux as a brand name but because of the commitment to the ideals behind freeware and the open source programming community.

## Information Age strategy: Napster and Gnutella?

Napster and Gnutella are programmes offered free of charge which enable file sharing over the Internet, and were specifically designed for sharing MP3 music files. Napster was designed in 1999 by Shawn Fanning, a student from Northeastern University in the US, as a file sharing system for music files. This single piece of freeware illustrated the power of the network society to an extent never seen before, creating an online community of users that numbered in the tens of millions within months of its first release on the Web (Kasaras 2002). Napster attempted to avoid the charge of breaking copyright law since it did not hold any music files on its own central server but merely enabled its users to share files stored on their own PCs, as such it was not directly responsible for music piracy. Despite this clever manoeuvre, the Recording Industry Association of America (RIAA) was somewhat successful in closing Napster down after it had collected a list of over 300,000 of Napster's users who had been identified as having transgressed the copyright laws using the Napster software (Barr 2000). The music industry has since been divided over the issue of whether Napster and similar programmes are good for music or not (Alderman 2001, Barbrook 2001).

One of the unintended effects of the RIAA's legal action was to stimulate further innovations in this kind of freeware, with even more sophisticated forms of file-sharing programme springing up, such as Gnutella, Aimster, Morpheus and Freenet (Barbrook 2001). For example, Gnutella provides a clear advance upon Napster in that it needs no central server to supply the file sharing capability. Instead, the PCs network directly with one another through their IP addresses. This provides a completely distributed network of users which constantly changes shape depending on who happens to be logged on at any particular moment. According to Barr (2000), 'Gnutella is definitely guerrilla warfare. There is no supply base, no central headquarters, no chain of command' (cnn.com/2000/tech/computing/05/12/mp3.guerillas.idg/index.html). Gnutella was originally developed as open source software released freely on the Web, and just like Linux and Napster it has spread like a virus since its inception.

Napster's major weakness was that although it exploited the power of the massive network of the Web, it was still based around a centralized database. The attempt to sidestep conflict with the music industry by not holding any music files on its own database ended in failure. The large corporations won their case and closed down Napster. Despite wining that particular battle, the question of whether they are winning the war is a somewhat different matter. Since the decline of Napster, more intelligent offspring have been spawned, such as Gnutella, which are becoming increasingly decentralized and hence more difficult to pin down and regulate. Another disagreeable consequence of this debacle for the RIAA and their tough stance on intellectual copyright is that their website has since become a popular target for hackers. This kind of retaliation against the behemoths of the music industry has become the subject of many reports in *Wired* magazine (January 2003).

The relationship between the nomadic lifestyle and music has very ancient roots. It is a curious coincidence that the earliest forms of exchange that emerged in primitive societies may well have centred upon the exchange of songs. According to Bruce Chatwin (1987), the central premise of the market mechanism, the idea of exchange, has its origins in the exchange of music. Chatwin's essay on the roots of nomadic aboriginal society observed that, 'Trade means friendship and co-operation; and for the Aboriginal the principal object of trade was song. Song, therefore, brought peace' (Chatwin 1987: 280). Napster can be seen as a direct extension of this nomadic tradition but raising it to a new level by exploiting modern communication technologies. Chatwin's essay also highlights the fact that behaviour of this kind is really a form of peace rather than a battle or war. Napster is not trying to fight a competing organization, but is fighting for a different way of life entirely.

Again, we can clearly see many of the principles of nomadic strategy at work in the case of Napster and its many offspring. The principle of speed and communication was notable in the incredibly rapid growth of the Napster community across the globe, from zero to millions in a matter of months. The principle of the idea was put to powerful effect by exploiting the important role that music plays in our lives. Music is enjoyed for its own sake but it is also strongly associated with the idea of friendship, and browsing other people's music collections online opened up new opportunities for developing new friendships. The principle of non-conflict was only partially successful in Napster's attempts to circumvent the intellectual copyright laws; however, it has succeeded in forcing the industry to rethink its distribution process and its relation with the artists on whom it depends.

## The Slow Food Society

Having been founded in 1986, the Slow Food Society was not directly inspired by the Internet, although it can be seen as a direct response to some of the homogenizing effects of globalizing organizations. The term 'Slow Food' is used to signify the society's opposition to the burgeoning fast food culture, stemming from the United States. This arose from a concern that the fast food culture, which has been developing apace with globalization, is gradually killing off much of the local cuisine and culture around the world. In particular, the aims of the society are to promote regional dishes made from produce native to that region. This is also in keeping with the ideal of preserving the local environment and native species of plant and animal. Indeed, the society has coined the term 'eco-gastronomes' to describe themselves (refer to its website for the *Slow Food Manifesto, International Statute*). The Slow Food Society was originally founded in the Piedmont region of Italy, and since 1989 has become an international movement. As of this year it has grown across five continents with over 60,000 members. These members are organized into a network of local grassroots groups called 'convivia', which provide feedback to the central offices in Bra in Piedmont. Each local group organizes local events and debates about Slow Food and the society also publishes books on Slow Food restaurants, and other issues concerning food, wine and culture.

This form of organization has adopted a strategy very much in keeping with the principles of nomadic strategy, as discussed previously. This strategy is not about conquering new territory but about resisting invasion by other outside forces (the fast food culture). This organization does not form a single disciplined corps, but is made up of a network of local activists, acting on a voluntary basis. The aims of this society are not simply about economics but concern the preservation of a way of life. In his study of the nomads of the steppes, Minns (1913) observed that even their methods of cookery were determined by their mobile way of life. In the case of Slow Food, the method of cookery is not itself determined by nomadism, although the strategy developed for the purposes of resistance is, to a large extent, related to the nomadic principles as described above. The Slow Food Society is particularly interesting because, even though it did not begin on the Internet, it has since embraced the Web as a powerful tool in promoting its message. In this respect it has behaved very much like the Zapatista rebels in Mexico, whose struggle is not confined to the Web like Napster and Linux, but nevertheless have used this communications technology very effectively.

## Innovators or parasites?

The use of nomadic strategy today often provokes a great deal of controversy. Napster exploded onto the scene in 1999, but disappeared almost as quickly. Opponents to Napster claimed that it damaged CD sales and therefore threatened the artists' livelihoods, a view that had the support from the Recording Industry Association of America, and artists as diverse as Metallica, Dr Dre and Neil Young. Other artists, such as Dave Stewart of the Eurythmics, took the opportunity to question the degree of control that large corporations already exercised over an artist's work (Barbrook 2001). Yet, even prior to Napster the recording industry recognized that the copying of popular albums onto tape had become widespread, if not common practice amongst their consumers. Some artists tried to use this new technology for copying and distribution to their benefit, such as the British band Radiohead. They released a part of their last two albums free on the Web, and thought it no coincidence that both of these albums went on to the number one slot in the sales charts in the United States, which was a first for the band in this important market. Alderman (2001) has pointed out that US CD sales had a bumper year in 2000, expanding by a further 4% a year when the music industry claimed to have been endangered by Napster. The most popular albums circulating on Napster were also those topping the US sales charts, suggesting that rather than being a threat to the industry, Napster was a superb marketing device.

Linux has also suffered at the hands of big business. In 1999, Microsoft intensified its information war against Linux by creating a new website under the heading of 'Linux Myths', challenging the reliability of this impertinent piece of software. Microsoft has produced numerous press releases that attempt to undermine perceptions of the quality of Linux by claiming that it has been developed by amateurs and hobbyists (Adelstein 2001). In an interview in June 2001, reported in the *Chicago Sun Times*, Steve Ballmer, the CEO of Microsoft, described Linux as 'a cancer' in the way it attached itself to intellectual property and then spread to other associated software by means of its distinctive form of copyright. Microsoft has even gone so far as to sue the distributors of some of the most successful releases of Linux, such as Lindows, for an alleged breach of intellectual copyright regarding the 'Windows' brand name. The information age nomads are accused of being thieves, in much the same way as nomadic tribes were condemned in ancient times. This attitude appears to have been a part of Western culture for a very long time indeed. Thinking from the perspective of the sedentary individual, the nomad may be considered to be a purely parasitic form of life, raiding cities and villages, and stealing food and people. Herodotus, for example, observed how the people of Athens distrusted the wandering nomads from the steppes, believing them to be uncivilized, barbaric and incapable of virtue. Today, this kind of accusation is reflected in the attacks levelled at online communities like Napster, who are supposed to steal from the mouths of the recording artists.

This critique of nomads as parasites is extremely partial. The simple fact is that even where they have run up against legal issues, nomadic organizations are often highly creative and innovative. Napster, for example, was the first online community that truly exploited the full potential of the Web, generating tens of millions of new members within its first few months. It demonstrated the enormous power of speed and connectivity in creating a new community orientated around music, and also the power of the Web as a means of marketing and distribution. The music corporation BMG Entertainment was one of the first to realize this fact, dropping its lawsuit against Napster in favour of an alliance. Even mainstream commentators, such as *The Economist* magazine, believe that

Napster has changed the music industry forever, and that big business will have to adapt to new modes of music distribution (*The Economist* November 2000). Linux also seems to have had a great impact on the computer industry in general. Red Hat is a company that thrives from offering support services for Linux. The Linux operating system has been adopted by big players in the industry, such as IBM and Hewlett Packard, and IBM has recently taken the opportunity to ally itself with Linux as part of its competitive strategy against Microsoft. It seems clear then, that in the cases of both Napster and Linux, business organizations are acting as parasites on the nomadic organizations rather than vice versa.

Communities like Linux are part of a wider group of programmers who have been broadly characterized as 'hackers'. As the Linux case shows, the objective of hacking is not necessarily theft but concerns an ethic dedicated to the free exchange of information. The group of researchers in the particle physics laboratory at CERN, who created the World Wide Web, were themselves animated by the hacker ethic, and they distributed their software free over the Internet (Himanen 2001). In contrast to a restrictive hierarchical architecture, their design using hypertext links was specifically created to facilitate the lateral transfer of information. No one owns the Internet and the direction it takes has no guiding committee; it is developed largely by an open community of hackers. The Web is by no means the only technological achievement that hackers have had a hand in developing. There is strong evidence to say that hackers were instrumental in the development of the Personal Computer (de Landa 1991, Himanen 2001). Prior to the invention of the PC, the main use of computer power was for batch processing business data. Even as late as 1977 industry leaders saw no other future, and the likes of Ken Olsen, co-founder of DEC, one of the largest computer manufacturers in the world at that time stated, 'There is no reason anyone would want a computer in their home' (cited in Himanen 2001: 187). In contrast to this sentiment, hackers were interested in increasing the level of interactivity with their computers. Two hackers, Steve Jobs and Steve Wozniak, created the first commercially successful PCs, the Apple I and the Apple II. In keeping with his hacker ideals and the pursuit of ever-greater levels of interactivity, Wozniak allowed the blueprints for these computers to be copied by other hackers. With the appearance of groups such as Linux and Napster, these hackers have adopted a nomadic strategy to fully utilize the power of the new distributed communications networks to further their ideals.

Deleuze and Guattari (1988) have written extensively on the significance of nomads as innovators throughout history. They point out that nomadic tribes were not just raiders but were innovators in areas such as metallurgy, jewellery and horsemanship. There is no straightforward relationship between the nomad and the city dweller, and there are as many examples of the city exploiting nomadic peoples and their ideas as there are of nomads stealing from the city. It is true that there were some nomadic peoples, like the Mongols and various Scythian tribes, who led successful raids on their settled neighbours, just as it is true that Napster fed off songs previously distributed by the music industry. Nomads are not dependent on the city and often live a life as far away, and as independently as possible, from city dwellers (Sellars 1998). For example, the ancient Scythian tribes were nomadic not because they needed to raid different towns and villages, but in response to the constant need to find new pastures upon which to graze their flocks. They moved like a pack, just as their herds moved to find new grazing. It would therefore be quite wrong to define the nomad simply in terms of the city dweller, in purely negative terms as 'uncivilized', since the nomads define their own way of life in terms largely independent from sedentary civilization.

Sellars has observed that nomads only become destructive when they run up against a rigid boundary that reduces their ability to flow (Sellars 1998: 79). Rather than being parasitic, nomadic people may be seen as great innovators. Their use of the horse for combat inspired various ancient armies to form mobile cavalry units of their own (Sellars 1998). Furthermore, ancient sources, such as Hesoid, have claimed that the nomads were the first to invent metallurgy for making weapons and tools. In his account of the history of the exchange economy, Marx attributed the invention of money to nomadic tribes. He provided two reasons for the impetus of this development: 'Nomadic peoples are the first to develop the money-form, because all their worldly possessions are in a moveable and therefore directly alienable form, and because their mode of life, by continually bringing them into contact with foreign communities, encourages the exchange of products' (Marx 1976: 183). Bruce Chatwin also observed that there is a common etymological root for the term nomad, from the Greek *nomos* and the Greek *nomisma* meaning 'current coin' (Chatwin 1987: 184). Hence the English term 'numismatics' – the study of coins. Chatwin observed that, 'Domesticated animals are "currency", "things that run" from the French courir. In fact, almost all our monetary expressions – capital, stock, pecuniary, chattel, sterling – perhaps even the idea of "growth" itself – have their origins in the pastoral world' (Chatwin 1987: 185). Deleuze and Guattari (1988: 368) have also pointed to medieval times when several journeymen's associations emerged that were made up of 'nomadic or itinerant bodies of the type formed by masons, carpenters, smiths, etc'. These itinerant professions were responsible for many of the great engineering and architectural feats of the period, particularly the appearance of the gothic cathedrals throughout Europe. These associations of itinerant workers can be seen as having strong parallels with the nomadic innovators of ancient times. Indeed, the likes of Napster, Linux, Slow Food and the Zapatistas are in many respects the information-age equivalent of such nomadic forms of organization.

## Towards a theory of information nomadism: smooth space

Network forms of organization are made up of distributed nodes or cells, located over a geographically dispersed area, and they have exploited modern communications technology to facilitate their networks. Some of the networks have adopted a sedentary approach to their construction, whereas others, of which the groups discussed above are exemplars, have adopted a more nomadic strategy to their development. The business guru, Michael Hammer, proclaimed the effectiveness of the sedentary type of network in his discussion of companies like Ford and their use of networks to 're-engineer the corporation' (Hammer 1990). Manuel Castells (1996) used the success of Cisco as his ideal of the 'network enterprise' because it has been the major provider of Internet technology during its explosive growth. Cisco also pioneered the use of the Web in dealing with its global network of customers, employees and suppliers. These networks tend to have what may be termed a 'sedentary' form of organization because they are designed to facilitate the regulated flow of information. This has been a particular concern of the literature on knowledge management, which for instance, proposes the use of 'knowledge maps' in order to regulate the movement and flow of information (Davenport and Prusak 1998). By such means as knowledge management programmes, intellectual copyright laws and access codes, organizations attempt to regulate the flow of information and to control its dissemination. Castells' own definition of the network enterprise is not one that allows information to travel in a nomadic fashion from any node in a network to any other, but one which covets its information: 'the network enterprise makes material the culture of the informational, global economy: it transforms signals into commodities by processing knowledge' (Castells 1996: 188). In the terms of Deleuze and Guattari, it is within such sedentary networks that deterritorialized flows of information and ideas become re-terroritorialized as commodities, and thereby become subject to ownership and control.

In contrast to these sedentary networks, in which the flows of information are domesticated, the kind of networks discussed in this chapter are distinctive because they all follow a nomadic strategy. This strategy is embodied in the eight principles outlined earlier in the section on Lawrence; however, these principles also form the basis of what might be called a theory of 'information nomadism'. To some extent, information nomadism was a major force driving the development of the Internet in the first place. In its prototypical form as the ARPANET, the Internet was designed as a communication network that could maintain its integrity even when a large number of its nodes had been damaged or destroyed (Castells 2001). It was designed to be independent from any central command centre, so that information could travel along any number of paths before being picked up by its intended recipient. This decentralized structure has been exploited and sees its ideal expression in the nomadic strategies that have been developed by the likes of Linux, Napster, the Zapatistas and the Slow Food Society. Firms within the software industry and the music industry have been surprised to discover threats to their business emerging not only from other corporations, but from diffuse communities of amateurs and hobbyists.

Deleuze and Guattari (1988) call sedentary networks 'striated space' that can be viewed as a whole from an external vantage point. This hierarchical point of view allows space to be divided up and subject to regulation (see the earlier critique of Weick). In contrast, nomadic space is a 'smooth space' which can only become known from within by means of legwork. Striated space is occupied by means of mapping it, counting it, parcelling it up and portioning it out. Smooth space is occupied by means of a vortical whirling movement. This distinction originally comes from the composer, Pierre Boulez, but Deleuze and Guattari (1998) show its application in other areas such as nomadic warfare, mathematics and even clothing. For example, the mathematical model they give for striated space is that of Euclidean geometry, but they compare smooth space with the intensive maps that can be generated from fractals. Smooth spaces tend to be difficult to regulate and police, like deserts, oceans, mountains and the steppe. Deleuze and Guattari (1988) observed that nomads are defined more by their geography than by their history, their ability to create and occupy smooth space. All of the cases discussed above are distinctive in their attempt to exploit the smooth spaces that may be created using new communication technologies such as the Internet. These are virtual spaces where information can spread in an unregulated, nomadic fashion. Indeed, the ideals behind freeware and copyleft are aimed at removing many of the barriers that block the free transmission and movement of information, and as such they are very similar to the idea of smooth space. In the terms of Deleuze and Guattari, these communities are attempts to deterritorialize the increasingly commodified spaces within which information and knowledge flows in our information-age societies.

The creation of smooth space is not always accompanied by a revolutionary movement. As noted above, the state itself often copies nomadic innovations. Deleuze and Guattari (1988) have observed that multinational corporations have also spread across the earth, deterritorializing its existing political and economic maps, and recreating territories in their own image. Likewise, innovations in warfare, such as the submarine and military aircraft, have transformed the sea and the air into a smooth space from which to attack the earth. Under certain conditions smooth space can in fact become part of the machinery of the state.

Nomadic strategy does seem to be particularly well suited to exploiting the new communications technologies, but it is by no means dependent upon them. For example, the Zapatistas, who Manuel Castells (1997) holds up as exemplars in the use of such technology, are largely comprised of poorly educated and poorly equipped recruits. The principles of nomadic strategy that were derived from T.E. Lawrence's guerrilla campaigns were also a low-tech affair, despite being orientated around the power of ideas and communications systems. These sentiments have many historical precedents, perhaps the most notable being Thomas Jefferson, who clearly expressed this kind of information nomadism in the following terms:

> That ideas should freely spread from one to another over the globe, for the moral and mutual instruction of man, and improvement of his condition, seems to have been peculiarly and benevolently designed by nature when she made them, like fire, expansible over all space, without lessening their density at any point, and like the air in which we breath and move, and have our physical being, incapable of confinement or exclusive appropriation.
>
> (quoted in *Wired Magazine* March 1994)

It is the attempt to exploit this peculiarly nomadic quality of information that unites groups as diverse as the Zapatistas, Linux, Napster and the Slow Food Society, and distinguishes their essential strategy. Information nomadism has many other forms, which may offer some interesting opportunities for future research. Efforts to striate the virtual space of the new communications systems are often subverted by disgruntled employees who may resort to hacking and leaving bugs or viruses on the system, or they may pass on sensitive information to rivals and competitors, or misinform consumers and shareholders (La Nuez and Jermier 1994). All sorts of resistance movements are coordinated on the Web, from the fuel-price protestors in the UK, to anti-capitalist demonstrators in Genoa, to the Zapatista rebels in Mexico. According to the RAND Corporation, this has led to the emergence of a new kind of warfare it has termed 'information warfare' (Arquilla and Ronfeldt 1997). Although some of these may be seen as infantile and misdirected acts of resistance, others are social movements that embody the protection of a community and a specific way of life. All of the cases discussed in this chapter concern communities that have adopted, in one form or another, the principles of nomadic strategy in a sincere attempt to protect and further their way of life, including organizations as diverse as the Zapatistas, Linux, Napster and Gnutella and even the Slow Food Society.

## Conclusions and connections

This chapter has outlined the general principles of a nomadic or guerrilla strategy. The nomadic wisdom of the Scythians was examined first to show that powerful strategies can be developed that do not assume the acquisition of property or new territories as their goal, but which can still sustain a community and a specific way of life. Similar ideas were at the root of

Lawrence's guerrilla war in the desert, and were summarized in the following set of principles: leadership, friendship, the idea, peace and non-conflict, the pack, calculation, speed and communication. However, the main thrust of this chapter has been to show that this powerful set of principles can be seen in all sorts of new social organizations emerging today, specifically in a number of cases including the Zapatista rebels in Mexico, the Linux and Napster Internet communities, and the Slow Food Society. These information age nomads have made significant innovations in the way we can organize ourselves, including new ideas (e.g. free property and Slow Food), new products (e.g. Linux and Napster), new forms of community based around file sharing, and new forms of resistance based around the concept of information warfare. Some of these organizations have sprung from virtual communities on the Internet whereas other have adapted information technologies to local struggles, as in the case of the Zapatistas and Slow Food. Each of these cases has adapted the principles of nomadic strategy differently to suit its aims and the conditions from which it has emerged. The Zapatistas, Linux and Napster may be seen as paradigmatic examples of information-age nomads, employing nomadic strategy to its full extent. The Slow Food Society is interesting because it only partially integrates nomadism into its strategy, but it highlights the fact that almost any organization can make use of this kind of strategy and the technology to facilitate it. This point will be discussed in more detail in a later chapter with particular reference the evolving concept of information warfare. For those of a military mind, one might say that the nomadic war machine is alive and well today just as it was in ancient times, or for those of a more pastoral persuasion, we can say that little rhizomes are sprouting up within our productive apparatus.

# 3 Mapping knowledge

## The new cartographers of organization

> The map served both to fix and segment the territory; to control it, to make claims of sovereignty and to package it for sale.
>
> (King 1996: 144)

> intellectual property rights are fast becoming a defining characteristic of our society.
>
> (Perelman 1998: 11)

> knowledge is not made for understanding; it is made for cutting.
>
> (Michel Foucault 1988)

The map has been suggested as a useful concept or metaphor in geography, philosophy, biology, and in the management sciences. All these different forms of information and knowledge become subject to visualization and are then mapped in order to facilitate organization and control. In the management literature, the map has been proposed as an important concept and has been discussed in relation to operational research (Eden *et al.* 1979), corporate strategy (Weick 1994) and knowledge management (Colville *et al.* 1999, Nonaka and Takeuchi 1995, Davenport and Prusak 1998). Ideas such as the 'information society' and 'knowledge management' are facilitated by the production of maps with which to regulate flows of information and capital.

This chapter explores the significance of the mapping metaphor for organizations today, with particular reference to the control of information and knowledge. Korzybsky (1958: 58) tells us that the map is not the territory. But if the map is not the territory, then what is it and how is it to be put to use? Maps need not be restricted to the features of physical geography, and may include man-made features such as monuments and castles, demographic and political features such as populations, territorial boundaries, and so on. Mapping is not simply a matter of symbolic representation but may be a part of a wider struggle concerning land ownership and property rights and the drawing of social and political boundaries. Colonial maps often portrayed native territories as empty spaces, where indigenous peoples were simply written out of existence (King 1996). The politics of map making was also unmistakable at the close of the Balkans conflict in the 1990s. When the different parties came to divide up Bosnia-Herzegovina, the map itself became a battleground where its definition and contestation became key to the resolution process (King 1996). Understanding mapping as a purely representational process is therefore a grossly over-simplistic view of the cartographic process.

The argument of this chapter is divided into three main sections: (i) the forces of capitalism that treat knowledge as a kind of territory that can be mapped, (ii) the points of resistance that resist such mapping, and (iii) a complete re-evaluation of the concept of mapping with respect to knowledge. The first section is devoted to the idea that knowledge can be transformed into a territory, which can be parcelled up and subject to control. The literature on knowledge management has argued that managers act as organizational mapmakers. This discussion leads to a number of problems associated with maintaining knowledge as a territory, which constitutes the second main section of the argument. Several difficulties may be encountered when treating knowledge as if it were a territory, including problems concerning the alienation of knowledge from the producer, the problem of isolating and controlling knowledge creation, and certain economic contradictions that are inherent in markets for information. The final section of the chapter re-evaluates the concept of the map, drawing on the work of Foucault and Deleuze and Guattari, which highlight two different concepts of mapping. The driving force behind knowledge management is that knowledge can be captured and subject to regulation, and is itself grounded in a representational model of knowledge. This representational account of knowledge can be contrasted with the idea that knowledge is best seen as a kind of labyrinth. The labyrinth concerns itself with those hidden tacit features of knowledge production and views the map as a kind of experiment with reality rather than as a representational model of reality.

## Knowledge: mapping a territory

One of the prime factors of economic production, whether manufacturing or otherwise, is information. Not only is information required for the control and organization of labour and machinery, it is itself a valuable resource that must be subject to control. Indeed, the establishment of rules for the control of information is one of the defining characteristics of the so-called information age. For the post-industrial economy, knowledge and information have been given increased prominence, particularly in terms of the value of expertise, credentials and intellectual property rights (Fuller 2001). The rather abstract practice of being able to 'manage knowledge' is itself becoming a new kind of expertise (Alvesson and Karreman 2001).

The economist Michael Perelman has observed that, 'The revolutionary aspect of the information age is the treatment of information as a commodity in ways that would have been unimaginable only a few decades ago' (Perelman 1998: 4). The value of knowledge management projects also tends to be assessed in terms of their contribution to the bottom line (McKinlay 2002, Xerox 2001). In order to measure the success of such projects, knowledge must be capable of being tracked and evaluated in financial terms. Methods for auditing intellectual capital are now being developed by some firms, such as Skandia, who now go as far as to include audits of their intellectual capital in their annual reports (Davenport *et al.* 1998). Social innovations, such as knowledge management programmes and the formulation of intellectual property rights, are perhaps more important in defining the information revolution than new technologies like the Internet or the Genome Project. The technological revolution has been driven and supported by various attempts to develop new regimes of control over knowledge and information, in terms of how it is produced, how one gains access to it, its format and availability, and how it can be used. The argument of this chapter is that the general form of this control involves the transformation of knowledge and information into a territory, and in particular a map or a series of maps.

Maps are generally employed for the purposes of orientating oneself in the world. They also serve to highlight different territorial boundaries, which may themselves be the subject of political conflict and are frequently subject to redefinition. King (1996) observed that advances in the technology of cartography were associated with the dawn of colonialism. Territories are often the sites of struggle, but this need not be confined to struggles over land ownership. Bogue (1999) suggests that music itself has territorial elements. Following ethnological studies, Bogue states that the concept of territory can extend to any defended area. The territory is itself subordinated to processes of territorialization and deterritorialization, whereby a territory is defined and its boundaries maintained. This definition is entirely commensurate with Knowledge Management (KM).

The literature on knowledge management tends to treat knowledge as a contested territory over which managers and organizations compete for possession. Karl Weick (1994) has proposed that managers are the mapmakers of organizations, where the territory being mapped is the collection of beliefs and ideas that are used to make sense of an organization. Weick's work explicitly follows Korzybski (1958), who proposed that language is a kind of map, a representational device for orientating oneself in the world. Korzybski further noted that 'A map is *not* the territory it represents, but, if correct, it has a *similar structure* to the territory, which accounts for its usefulness' (Korzybski 1958: 58). Weick draws on this idea, but adapts it by saying that, because of the interpretative nature of social organizations, managers can create their own territories. The real task of management, according to Weick, is this process of mapping, and more specifically, territorializing the ideas and beliefs of those who work within their organization.

## Knowledge management as cartography

In the management literature, mapping has been proposed as a primary task of management (Weick 1994, Colville *et al.* 1999). Mapping is now the explicit mechanism that managers are being urged to adopt by the advocates of knowledge management (Davenport 1995, Davenport and Prusak 1998). Almost all the large consulting firms claim to be doing knowledge management of some kind, including Accenture, KPMG, Ernst and Young, and McKinsey and Company. Xerox (2001) reports that knowledge management is being widely adopted in companies throughout Europe. Its academic credentials ought to be beyond dispute and have been established in many ways, for instance universities have begun to run courses in KM and have established professorial chairs and research centres. The most credible form of legitimation, at least in academic terms, is the torrent of articles on KM appearing in refereed journals from the early 1990s onward, including the *Harvard Business Review*, the *Sloan Management Review* and *Organization Science* (Nonaka 1991, 1994, Davenport *et al.* 1998). To the uninitiated it might appear that KM stands on firm theoretical foundations, but there are dissenters who challenge the very basis of much of this work. For example, one recent overview of KM has stated that 'the majority of even the "academic" literature on Knowledge Management was, frankly, so simplistic that attempting to position our discussion *within* that literature was not an option' (Hull 1999: 416). Much of the pioneering work in KM, such as that of Nonaka and Takeuchi (1995) is devoted to the regulation and control of explicit and tacit forms of knowledge within organizations. However, some reviewers have claimed that the entire edifice of Nonaka and Takeuchi's work rests on a simple mistake – the possibility of converting tacit knowledge into explicit knowledge (Tsoukas and Vladimirou 2001). Others have been more generous, suggesting that although tacit knowledge cannot be converted directly into explicit knowledge, the interaction between these different types of knowledge can create new knowledge (Cook and Brown 1999).

Knowledge management has taken off even if no one seems to know exactly what it is. This may not be harmful in itself and may leave room for a certain amount of experimentation. One may know what 'KM' is insofar as it involves the re-packaging of existing ideas, such as the learning organization (the approach of Nonaka and Takeuchi) or information systems (the approach of Davenport and Prusak). The problem is, as the following section will show, that not only is the general concept of KM unclear, but that much of the pioneering theoretical work on 'knowledge management' is deeply flawed. Despite this, knowledge management has proven successful both in academic terms in developing research posts, courses, publications, and in business terms in its adoption by the big consulting firms and the willingness of companies to pay for knowledge management services.

It is possible that the academic arena has played a largely unreflective role in legitimating and supporting a theoretically weak concept. As an academic, one might be surprised at the casual way in which this has been received given that knowledge is the name of our game, so to speak. In post-industrial economies, the status of knowledge and its validation is not simply a matter of establishing its truth, but is crucial to establishing its exchange value. According to a recent OECD report, answers to such questions form the basis of much of our economy where, 'more than 50 percent of Gross National Product in the major OECD economies are now knowledge based...' (Perelman 1998: 14). As these words are being written the crucial role of auditing in the information economy has become all too clear, with the recent disappearance of many hundreds of billions of dollars from the value of shares on the US stock exchange. Auditors, such as the accountancy firm Arthur Andersen, have been validating misleading company accounts for some years, and they are not the only ones. Arthur Andersen is no longer in business as a result of legal proceedings against it, and other large financial institutions such as Citigroup and Merrill Lynch have been fined hundreds of millions of dollars for providing misleading advice to their customers. Juvenal's sardonic question in *Satires* appears to be particularly apt: 'Who shall guard the guardians?' Auditing plays a special function in the post-industrial economy for many reasons, not simply when asking how much a given company may be worth, but when questioning the integrity and value of its stock of knowledge and information. This seemingly intractable problem will be explained later, when we consider how markets for information work. We will now turn to a more detailed discussion of the theoretical basis of knowledge management and the treatment of knowledge as a territory.

The mapping metaphor is frequently used in the literature on knowledge management (see for example, Nonaka and Takeuchi 1995, Davenport and Prusak 1998, Hull 1999). Knowledge management gurus recommend that organizations develop an explicit 'knowledge map' to utilize effectively the organization's intellectual resources. Many large organizations have begun creating their own knowledge maps, including Monsanto, Hoffman-Laroche, Health Canada and the US Army, a trend that appears to be on the increase. Much work in the knowledge management literature does not question the theoretical basis of knowledge management, despite providing sound empirical studies of corporate KM programmes (e.g. Xerox 2001, McKinlay 2002). These studies leave the concept of tacit knowledge largely unquestioned, although this is the basis upon which much of the initial KM theory was grounded. In practice, knowledge management programmes are largely concerned with knowledge that is explicit and relatively easy to record. McKinlay (2002) describes an extensive knowledge management programme within a large pharmaceutical company that uses a variety of means for processing the collective knowledge of employees in the form of 'lessons learned', 'ad hoc local solutions' and 'horizontal conversations' among other things. The element of knowledge that is the subject of such programmes is really the explicit, usually linguistic, dimension. The KM literature tends to treat knowledge as a possession, as part of the system of exchange, and an element in the global commodity markets. This is clearly reflected in popular texts on knowledge management such as those of Davenport and Prusak (1998) and Nonaka and Takeuchi (1995), and various case studies which discuss the kinds of technologies used in the implementation of knowledge management programmes (Xerox 2001, McKinlay 2002).

Both the theory and the practice of knowledge management are dominated by a concern for explicit knowledge, or the attempt to make knowledge explicit. For example, the work of Nonaka and Takeuchi (1995: 230–231) is concerned with tacit knowledge insofar as it can be made explicit and therefore subjected to control. In contrast, Davenport and Prusak (1998) suggest that when tacit knowledge cannot be 'converted' into explicit knowledge, it can be transferred between people by means of mentoring and apprenticeships. Knowledge maps should be drawn up which can serve as 'a guide, not a repository' (Davenport and Prusak: 72). Davenport and Prusak (1998: 81) maintain that 'knowledge maps' should not replace people but should act as a pointer to those with the relevant knowledge and expertise. Whereas Nonaka and Takeuchi (1995) claim that their form of knowledge management is 'universal', Davenport and Prusak are more sensitive to the problems inherent in the globalization of business. They point out that there might be some conflict between local knowledge and knowledge that can be easily shared between different cultures. To preserve local knowledge, only the most essential terms should be standardized. Despite some sensitivity to the limitations of knowledge management they still maintain that, 'As difficult as it is to codify tacit knowledge, its substantial value makes it worth the effort...' (Davenport and Prusak 1998: 81). Accordingly, they make the same error for which Nonaka and Takeuchi have been criticized, the presumption that one can somehow 'convert' tacit knowledge into explicit knowledge. There are a number of examples that exemplify their enthusiasm for ever-increasing levels of formalization, the most absurd of which is probably that of Microsoft. Microsoft has developed a knowledge map that rates employees according to their 'knowledge competencies'. There are 137 implicit and 200 explicit knowledge competencies, and each one of these is itself broken down further into four more levels (Davenport and Prusak 1998). The stated aim is to render the 'knowledge competency' of employees both explicit and measurable, providing information that can easily be interrogated using a database. Davenport and Prusak concede the importance of tacit knowledge, but it merits only a few pages of discussion in the entire work.

The concept of mapping knowledge should not simply be understood as a metaphor for knowledge, because it is embodied in the very technologies being used to accomplish this transformation (Elmer 1999, Hull 1999). Hull's overview of contemporary knowledge management projects concluded that, 'various "mapping" techniques we found were quite varied, but were more concerned, metaphorically and sometimes literally, with visualizing the domain in question' (Hull 1999: 421). Elmer makes the point that there appears to be a convergence between mapping techniques such as geographical information systems and marketing techniques, which draw on a variety of sources of data including geographical data, demographic data, psychographic data and consumer behaviour data. These sources of data form part of an indexical apparatus which can be used to underpin the coding of data to produce a map for the purposes of market segmentation. Elmer perceives a direct analogy between the application of such mapping techniques and the Panoptic technology described by Michel Foucault (1977). Elmer's analysis of the use of mapping techniques today concludes, 'To map is . . . not only to document a governmentalized space, it is, moreover, itself an ongoing attempt at governing space and lifestyle' (Elmer 1999: 59). Visibility is not immediately concerned with the body, as in the case with the traditional Panopticon, but concerns the flows of information generated about people and their activities. This is symptomatic of a change in power relations, where the flows of knowledge and information are increasingly becoming the object of power, rather than just the physical exertions of the disciplined body (Deleuze 1995). Techniques for mapping knowledge and information are a key feature of the attempt to regulate and control the flow of knowledge. These techniques do not just apply to the government of lifestyles but to the government of life itself, as we shall see in the following section on the use of mapping techniques in the field of biotechnology to facilitate both scientific inquiry and the establishment of intellectual property rights associated with biotechnology.

## Mapping the body: intellectual property and biosocial maps

The concept of 'intellectual property' and the increasing commodification of knowledge are currently being transformed by developments in the biotechnology industry. Plant life has been subject to patent since the 1930s, but in 1980 the patent laws were extended to cover other life forms. This began when Ananda Chakrabarty engineered a bacterium for General Electric that digested a component of crude oil and could be used as means for clearing up harmful oil slicks in the sea. This genetically modified bacterium was deemed to be a human invention and, as such, subject to patent. Prior to these biotechnological innovations, the natural world had been considered as publicly available for use by all. The 1980 legal decision dramatically challenges this supposition and may have profound implications for the development of any societies trading with the US. As Rabinow (1996b: 21) explains: 'The Chakrabarty decision validated a new dimension in the place "nature" would have in both the scientific and cultural world'. Nature itself is now a technological product, and in Rabinow's terms we have entered a period of biosociality, where 'nature will be known and remade through technique and will finally become artificial just as culture becomes natural' (Rabinow 1996a: 99).

The first animal to be patented was to come in 1988 when Harvard College patented the OncoMouse, licensed to Du Pont. The OncoMouse is a mouse that has been genetically engineered for its predisposition for breast cancer. Life is beginning to be transformed into intellectual property as the limits of capitalism continue to expand. The human genome is on the way to becoming the next biosocial threshold to be crossed, and perhaps the most profound. The genome is gradually being transformed into a commodity fetish (Rabinow 1996a, Harraway 1997). Harraway discusses how the human genome has been portrayed in terms of a map in both the popular and scientific literature. According to Harraway, genes should not be understood as a representational map of the body or as 'master molecules', since they are one element in a wider system of intervention, they 'are always part of an interactional system' (Harraway 1997: 145). The human genome is less a biological reality than it is a 'fictive composite', or the portion of DNA that is presumed to be held in common by all persons (Flower and Heath 1993: 29). Rabinow discusses this problem in relation to the concept of normalization. He invites us to ask, 'Whose genome is it? Obviously not everyone has exactly the same genes...' (Rabinow 1996a: 97). Rabinow observes that scientific discussion of this question is virtually non-existent.[1] Given current mapping techniques, the easiest genome to map would be the one containing the most abnormal genes. As such, 'the pathological would be the path to the norm' (Rabinow 1996a: 97). Not for the first time, an important social or biosocial question may ultimately be resolved in terms of economic and technological feasibility.

The ability to copyright gene sequences has opened up a new biological and business frontier called 'bioprospecting'. Bioprospecting is the term used when large, usually Western, pharmaceutical companies codify and patent the gene sequences of plants and animals from areas across the globe, often Third World countries. Perhaps the most controversial form of bioprospecting is the patenting of gene sequences from indigenous human communities who are known to possess certain inherited qualities. Concerns have been expressed that such bioprospecting will become yet another division between the First and Third Worlds, to the extent that the companies of the developed world might even come to own the genetic inheritance of the peoples of the less developed world (Polster 2001, Ostergard *et al.* 2001). Environmental pressure groups like Greenpeace have begun to take this issue very seriously, as demonstrated in their current legal case against Monsanto and its patent on a strain of wheat popular for making bread in large parts of India (*The Guardian* 31 January 2004). Greenpeace argues that such patents ignore the creative efforts that previous generations of local farmers had put into refining the Nap Hal wheat variety, well before companies like Monsanto and Unilever appeared on the scene.

The biosocial field is expanding in a number of directions at present, where organizations as diverse as insurance companies and crime agencies have an interest in gaining access to records of people's DNA. Genes are thus becoming an important part of the administrative mechanism of both private and public organizations. It is not too difficult to imagine a future society where social divisions are manifested as much in the quality of a person's genetic inheritance as they are currently in terms of their wealth or the kind of work they do.[2] It is clear from recent innovations in intellectual property and bioprospecting that a biosocial map of inequality is becoming directly connected to the biotechnological map of the genome. Mapping is thus becoming a crucial feature of the post-industrial marketplace of ideas and it should be of no surprise that the managerial class has taken up this metaphor as part of the development of many of its latest ideas and methods. However, the various techniques that are being developed for mapping knowledge and rendering it a territory are by no means unproblematic, and it is to such problems that we shall now turn.

## Problems and contradictions of mapping knowledge

> The possibility of the intelligent direction of production expands in one direction because it vanishes in many others.
>
> (Marx 1976: 482)

Knowledge and information are increasingly being treated as a territory, where managers are acting as cartographers in order to facilitate the control of valued knowledge. One of the driving forces behind the mapping of knowledge and information is that, when seen as a scarce resource, it can become a very valuable commodity. In fact, the concept of the commodity and that of information share many important characteristics. Both are defined in terms of their immaterial nature. The father of cybernetics, Norbert Wiener (1950) defined information in terms of a pattern, which describes the underlying order of the world abstracted from its physical dimensions. In a similar fashion, Marx explained the commodity in the following terms 'as soon it [a material object] emerges as a commodity, it changes into a thing which transcends sensuousness' (Marx 1976: 163). Information and the commodity are both abstractions that serve as a locus of exchange. Wiener, for instance, defined information as a locus of exchange between the various parts of a cybernetic system, between man and machine. The commodity is also a basis for exchange, and defines all economic exchange under capitalism, where the commodity can be defined by its 'exchange value' in terms of a quantity of money.

Despite the apparent similarities between information and the commodity, Wiener warned that there are also significant differences. Information is subject to the second law of thermodynamics and as such it is not conserved when it is transmitted. In his own words, 'Property in information suffers from the necessary disadvantage that a piece of information, in order to contribute to the general information of the community, must say something substantially different from the community's previous stock of information' (Wiener 1950: 132). Put simply, the more times a piece of information is exchanged, the less it will be worth. This important limitation on the role of information as a commodity has also been analysed by economists such as Arrow (1984) and Perelman (1998), whose work will be discussed in greater detail below. In fact, there are many points of resistance that make it very difficult to map knowledge and transform it into a commodity. These points of resistance have featured as major issues of contention in the literature on knowledge management, especially surrounding the debate over the possibility of converting tacit knowledge into explicit knowledge (Cook and Brown 1999, Tsoukas and Vladimirou 2001). This is not the only sticking point, as will be explained below under the following headings: (i) creativity, (ii) alienation, (iii) scarce resources, and (iv) anti-markets.

### *Creativity*

It is clear from the above discussion that the weakest part of the knowledge management literature is that element which deals with creativity. Nonaka and Takeuchi (1995) spend a great deal of space engaged in questionable argumentation concerning the transformation of tacit knowledge into explicit knowledge in order to unravel the knowledge creation process, yet one needs only turn to the final page of their manifesto, where they humbly claim that the 'knowledge-creating process is no longer an enigma' (Nonaka and Takeuchi 1995: 246). Not surprisingly for a book that claims to have mastered the knowledge creation process, Nonaka and Takeuchi devote much time to those aspects of knowledge that defy formalization and regulation. This is quite clear in their lengthy discussions of tacit knowledge: 'The quintessential knowledge-creation process takes place when tacit knowledge is converted into explicit knowledge' (Nonaka and Takeuchi 1995: 230–231). Tacit knowledge, then, is not really knowledge at all, and it must be made explicit before it can claim that title. It must become converted into a 'formal and systematic language', one that is visible and subject to commodification and control. But how is this transformation to be achieved? Only when this esoteric process is made clear can it be said to have cut through the labyrinth of creation. According to Nonaka and Takeuchi this process is much simpler than one might have suspected, claiming that, 'Since tacit knowledge is inexpressible, metaphors and analogies serve as the means of expression' (Nonaka and Takeuchi 1995: 231). This statement is plainly contradictory, it begins by conceding that tacit knowledge is inexpressible, and yet it continues by attempting to explain how it can be expressed. The method Nonaka offers for the conversion of tacit into explicit knowledge is the dialectic, 'Dialogue directly facilitates this process by activating externalization at individual levels' (Nonaka 1994: 25). However, the problem of offering dialogue as the conversion process is that the dialectic is itself an explicit form of knowledge, and deals solely with what is expressible. It is by no means clear that the dialectic converts tacit knowledge into explicit because the dialectic is concerned only with terms that are already explicit. When followed step by step, this proposed knowledge creation process evaporates before our very eyes. Cook and Brown (1999) also locate the knowledge creation process in the interaction between tacit and explicit knowledge, with the added provisos that, (a) it is not converted but actually created and, (b) knowledge should be further divided into that held by groups or individuals. According to the authors, the knowledge creation process is a 'generative dance' between existing forms of tacit, explicit, individual and group knowledge. However, the precise nature of Cook and Brown's generative dance is not itself explicit and the knowledge creation process remains an enigma.

In contrast, Davenport's work does not claim to reveal the creative process to us, and leaves tacit knowledge where it is, embodied in people (Davenport and Prusak 1998). Instead, Davenport recommends the use of a map which points the inquirer to the appropriate expert, when needed. In practice, most knowledge management projects are concerned primarily with knowledge sharing, and the issue of creativity occupied a relatively minor role (Xerox 2001). The research of McKinlay (2002) is one of the few attempts to explore the contradiction between creativity and control. McKinlay explains a fairly adventurous and open-ended knowledge management project in a large pharmaceutical firm where a great deal of leeway was permitted to those engaged in the project to encourage creativity. However, there were institutional pressures against allowing such flexibility, because it was unclear how the performance of such a project could be measured or seen to pay off. It would seem then, that attempts to control and manage knowledge reach their limits when confronted by whatever processes knowledge comes to be created, where the impositions of tighter controls may even deter this very process.

### *Alienation*

One should not be surprised to discover that the increasing commodification of knowledge goes hand in hand with an intensification of alienation. The commodification of knowledge depends to a large extent on the alienation of knowledge from the producer, as indeed all commodities depend upon this very process. Fuller (2001) categorizes post-industrial work in order of increasing alienation: expertise, credentials and intellectual property. Expertise is the least alienated because it is somewhat indistinguishable from the character of the expert, credentials are something a person may acquire, and intellectual property is something a person may produce in a form independent of their person.

There is a sense in which knowledge management, as a function of management, is directly concerned with alienation. Frederick Taylor had a form of knowledge management at the heart of his idea of scientific management: 'The duty of gathering in of all this great mass of traditional knowledge and recording it, tabulating it, and in many cases, finally reducing it to laws, rules, and even to mathematical formulae, is voluntarily assumed by the scientific managers' (Taylor quoted in Perelman 1998: 40). However, Taylor's idea can be differentiated from the current knowledge management fad because his approach relied on removing decision-making abilities from the manual labourer. In contrast to clear-cut Taylorism, knowledge management is about ensuring the control of knowledge and ideas without removing the need for workers to think about the work in which they are engaged. This is one reason why knowledge management texts focus so strongly on the desire to make tacit knowledge explicit, because the knowledge worker is more autonomous and more dangerous due to their expert tacit knowledge. In order to exercise control over such workers and the contents of their brains, their knowledge must be mapped by means such as knowledge management programmes or the imposition of intellectual property rights. This, of course, requires an intensification of the degree of alienation between knowledge and producer.

The Xerox survey of knowledge management programmes throughout Europe concluded that, 'Increased staff knowledge, satisfaction and motivation are considered to be *the least important* ... objectives' (Xerox 2001: 13, emphasis in original). The same survey found that one of the major impediments to successful knowledge management programmes is the reluctance of people to share their knowledge. Sharing knowledge is a key component of knowledge management theory and practice. Some firms attempt to overcome any reluctance amongst their employees by offering financial rewards to those that are considered good at sharing their knowledge (Davenport *et al.* 1998). Although the literature on KM emphasizes the importance of sharing knowledge, this has very strict boundaries. In fact, unauthorized knowledge sharing can be used as a form of worker resistance where disgruntled employees may pass on sensitive information to outside interests or circulate misinformation within their own organization (La Nuez and Jermier 1994). In spite of the rhetoric, the strictures imposed by companies on knowledge sharing are expanding into all aspects of social life, for instance IBM has recently sacked one of its employees, Virginia Rulon-Miller, for dating someone who worked for a competitor (Perelman 1998: 84). Unauthorized sharing, or even its possibility, is strictly forbidden and to be guarded against. There is also evidence that the scientific community is becoming increasingly secretive with respect to its work, because of the strictures imposed by corporate sponsors (Marshall 1997). Alienation is taking on a new dimension as it places itself at the core of the processes of knowledge production.

### *Is information a scarce resource?*

There are also many problems with treating information as a commodity from a purely economic point of view. These problems have been considered in some detail by economists such as Perelman (1998) and Nobel Prize winner Kenneth Arrow (1984). Most commodities can be easily appropriated by the owner, but this is not the case for information 'because an individual who has some can never lose it by transmitting it' (Arrow 1984: 142). Information is inappropriable and it can be treated as a commodity 'only to a limited extent' (Arrow 1984: 142). As such, the whole idea that the market will lead to an efficient allocation of resources becomes problematic.

Information has an exceptional status amongst the many commodities that circulate the global markets because one of its most unusual qualities is that when it is consumed, it does not decrease. In fact, precisely the opposite is the case, because information actually spreads and increases as it is consumed. Perelman (1998) has observed that the spread of information can directly facilitate the production of yet more of it. In economic terminology information is 'non-rivalrous', which means that it is not exhausted in the act of consuming it. If this is the case, then traditional economic explanations of how the market for information works simply cannot function. This was pointed out by Arrow, who observed that 'Information is not scarce in the same sense as other commodities are . . . Patents and copyrights are social innovations designed to create artificial scarcities where none exist naturally' (Arrow quoted in Perelman 1998: 87). The market for information is successful only to the extent that barriers can be developed and erected to exclude unwanted access to the information on sale. So much for the myth that the information age is ushering in a utopia of information for all. The information age has become obsessed with making airtight containers for information, and hence the current interest in passwords, PIN numbers, encryption, intellectual copyright, intranet firewalls and so on. The growth in communications networks has simultaneously been accompanied by a revolution in the technologies that protect and prevent the flow of information.

What easily applies to information technology cannot be so easily applied to the human brain or our casual conversations; consider the absurd example of IBM's sacking of Virginia Rulon-Miller. Arrow summarizes this problem succinctly by stating that whatever the container in which information is kept, 'if discovered, it is likely to leak in some form' (Arrow 1984: 202). The privatization of informational goods creates other problems. The cost of gaining credentials, expertise, and information in the form of reports or academic journals is rapidly increasing (Perelman 1998). This is leading to a widening gulf between the information haves and have-nots. Fuller has observed a comparable trend at the heart of the education system itself: 'from being a principle of empowerment, credentials have now become indirect markers of exclusion' (Fuller 2001: 186). We may be led to the strange conclusion that the market functions by creating scarcity rather than the other way around.

### *The creation of information anti-markets*

Arrow highlighted yet another flaw in the idea of information markets in the following statement, 'We can possess an informational product merely by learning about it. Consequently, the owner strives to keep the information as secret as possible, preventing us from shopping for information in an informed manner' (Arrow quoted in Perelman 1998: 89). This clearly suggests that although information can be bought and sold, it will not lead to an efficient allocation of resources. Arrow's own economic analysis forces him to conclude that, given that the cost of transmission is so low, the marginal cost of information is effectively zero and so it cannot command a price (Arrow 1984).

Arrow frequently uses the example of expert knowledge in his analysis of information as a commodity. For instance, the patient is completely reliant upon the doctor for expert knowledge of the state of their health, and can pay large sums of money for this advice. This relation of dependence is such that it may not be possible for the patient themselves to determine whether or not they really need the expensive treatments they may have been prescribed. As such, Arrow states that the marketplace is not sufficient to determine the price of such expert knowledge but instead a high degree of trust in the expert is required, which might also be supported by a code of professional ethics. Arrow can be seen to be following in Juvenal's footsteps by stating that, ultimately, there is no one guarding the guardians, and at some level they must be trusted to regulate themselves (Arrow 1984: 150). When this system is flouted, the economic market for information begins to disintegrate. Indeed, the current stock market crisis in which Arthur Andersen has played such a prominent role may be seen to be a perfect illustration of this problem. Although some of the more critically minded economists may be puzzled at the existence of such contradictions in information-based economies, one should bear in mind that this does not imply that such a contradictory system cannot work. Indeed, an enduring feature of the capitalist system has been its ability to absorb contradictions without the need to resolve them (Hardt and Negri 2000).

So far, this chapter has made two major arguments, the first of which is that knowledge and information are becoming increasingly treated as a territory subject to mapping techniques which facilitate its ownership and control. The second major argument is that treating knowledge in such a territorial fashion has a number of serious problems: its hidden tacit dimensions resist formalization and control, and it suffers from inherent economic contradictions. In the following section we will discuss the tensions outlined in the four points discussed above in terms of alternative approaches to knowledge management. This will be approached in two stages, first by looking at the creative process in terms of a re-evaluation of the concept of the map, and secondly, by discussing markets for knowledge and information in terms of an intellectual commons.

## Re-evaluating knowledge management: post-structural alternatives

### *Models of representation and labyrinths of creation*

In the following argument, a distinction will be made between two different uses of the map. On one hand, a map may be used as a model to represent reality, whereas, on the other hand, the map becomes a labyrinth. A short story by Italo Calvino may help to illustrate this distinction. In his short story, *The Count of Monte Christo*, Italo Calvino re-tells Dumas' tale of unfair imprisonment and eventual escape but with an added twist. In Calvino's tale, a heroic escape cannot be taken for granted. Calvino explores the different ways in which the two inmates approach their problem; in his own words:

> On the basis of Faria's mistakes Dantes tries to draw a map of the castle. While Faria, by the sheer number of his attempts, comes close to achieving the perfect escape, Dantes moves toward imagining the perfect prison – the one from which no escape is possible.
>
> (Lecture in 1967 reprinted in Calvino 1997: 26–27)

Dantes believes that if he draws a map of the perfect prison, either it is real and he cannot escape, or it is superior to his own confines in which case he just has to find the point where they do not coincide to find a means of escape. The map serves less as a model of reality, than it does a process of experimentation with the unknown and, in this case, a possible means of escape.

To unravel the labyrinth is to destroy its power, but at what point does this happen? Calvino stated that as soon as one sees the map simply as a model of the world one ceases the attempt to understand the world. There is a striking similarity between Calvino's story and Foucault's descriptions of the Panopticon. Foucault (1977) used the Panopticon to map out not just prisons but society as a whole, in schools, barracks, factories and so on. This ideal prison provides a map of the entire social field. For Foucault the problem was not so much that prisons are everywhere, but that we seem incapable of thinking otherwise. It is in this sense the Deleuze (1988) described Foucault as a 'new cartographer'. Like Dantes, he draws a map to better aid escape.

Deleuze and Guattari (1988) have suggested that maps should be seen as a kind of experimentation. They criticize the representational understanding of maps, which they call the 'tracing' or 'model' (Deleuze 1998, Deleuze and Guattari 1988). The tracing is never a tracing of the map, since the map concerns itself with what avoids representation. Whereas the tracing or model presupposes the existence of a circumscribed unchanging reality, the map is entirely concerned with making new connections and transformations. The map is 'entirely oriented toward an experimentation with the real' (Deleuze and Guattari 1988: 12). Experimentation is not a controlled process, it is not an experiment on a model; instead it takes place in contact with the unknown, it takes place when alien forces meet. Kafka did not write about his trial to find us guilty, neither did Foucault design his Panopticon to trap us. These map makers do not tell us where we are, they tell us that we are in a labyrinth and open up new possibilities for living. Deleuze described the map as a rhizome or a labyrinth of creation and transformation where, 'The labyrinth is what leads us to being, the only being is that of becoming, the only being is that of the labyrinth itself' (Deleuze 1983: 188). The map is not a model of the labyrinth but is a map of transformations and becomings (Deleuze 1998).

In his *Reflections on The Name of the Rose* Umberto Eco distinguishes between three different understandings of the labyrinth (Eco 1984). The first sense is the classical labyrinth, which continuously folds back and forth on itself, but has only a single path that inevitably leads to its centre. This is the mythical labyrinth of the Minotaur that Daedalus built for King Minos of Crete. The second kind of labyrinth is the maze, which contains many paths and dead ends but through perseverance one may reach its centre. This might be called the modern labyrinth (Jaskolski 1997). Finally, Eco writes of a post-modern labyrinth, which he explicitly compares with what Deleuze, G. and Guattari, F. (1988) called the rhizome. Eco explains that this labyrinth is not a maze or a library, but the story itself; it is 'a story of labyrinths, not only spatial labyrinths' (Eco 1984: 58). The labyrinth itself should not be seen as an architectural prison. As in Leibniz's perfect encyclopaedia, we engage with knowledge as we follow where it takes us. Inside the labyrinth, information and knowledge is simply not 'manageable' from a position of exteriority. Jaskolski's (1997) study of the labyrinth offers evidence to suggest that the original Cretan labyrinth was not really a structure at all, but was a dance that Theseus performed for Ariadne. In this dance, Theseus was connected by Ariadne's thread to a chain of other dancers. The labyrinth is the dance and as such it does not represent movement, it embodies movement.

Ariadne's other lover was the god Dionysus, who was the god of music, dance and the labyrinth. Music and dance are all aspects of the unconscious powers of the body. In modern terminology, the god of the labyrinth should be understood as the god of the unconscious (Deleuze 1988). The labyrinth describes the dance, where the dancers are lost in the movement of the dance, serving as a metaphor for the embodied and unconscious nature of thought and action. Perhaps, it is here that we may look for the wellspring of knowledge, below the surface of consciousness. This is precisely what Polanyi's (1967) work investigated in his original discussion of tacit knowledge.

Polanyi did not dispute the value of scientific method in rigorous, systematic research. However, he observed that scientific discovery also proceeds by intuition based on the tacit knowledge of scientists, by means of which information can be synthesized and existing problems recast. It is this work on tacit knowledge that inspired the later theorists of knowledge management, but the significance Polanyi gave to the body and its unconscious forces have been largely neglected. There is an element of this approach to knowledge in the work of Cook and Brown (1999), who have criticized existing approaches to knowledge management for being too focused on reducing knowledge to the status of a possession. Cook and Brown focus less on knowledge as an entity and more on the process of knowing, which they say can only be truly appreciated in terms of our bodily interaction with the world. Polanyi's work is undoubtedly an important contribution to the philosophy of science, but he is by no means the only thinker to move away from an 'epistemology of possession'. Many thinkers may be drawn on whose work moves us away from an epistemology of possession in both the natural and social sciences, including authors as diverse as Nietzsche, Bergson, Heidegger, Foucault and Deleuze. Such thinkers do not treat knowledge as a straightforward representational tool, but as a way of relating to the world, and an essential mediator in defining what it means to be human. It is not simply about rational problem solving, but about how we pose problems in the first place. Innovation is not primarily a matter of rational problem solving, but of creating the terms in which a problem is expressed. Deleuze (1991: 16) has observed that 'the history of man, from the theoretical as much as from the practical point of view is that of the construction of problems'. This process of problem construction is common to both the arts and the sciences, and in each case requires the use of both intuition and intelligence. Furthermore, it is by means of this process of constructing its own problems that humankind becomes conscious of its freedom. Mapping therefore is not simply a straightforward process of representation, but is also a process of posing problems.

### *Knowledge as commodity and knowledge as common good*

There are some initiatives being developed that treat knowledge as a territory, but still pose an alternative to its increasing commodification. These initiatives are aimed at breaking down the financial barriers to its free dissemination and movement. There are many proponents for treating knowledge as an intellectual commons, which is open for general usage as a public good. This idea is being entertained in many different areas including international relations (Hardt and Negri 2000), economics (Perelman 1998), computing (Moody 2001), and sociology (Polster 2001). Even a former President, Thomas Jefferson, advocated the intellectual commons, believing that ideas were, 'incapable of confinement or exclusive appropriation' (quoted in *Wired Magazine* March 1994).

The creation of an intellectual commons should not be understood merely as a lack of knowledge management, but as an alternative form of knowledge management in its own right. There are a number of high profile cases of organizations that have consciously adopted this form of knowledge management, such as the Massachusetts Institute of Technology, who are making their course materials freely available on the Web in an initiative called *OpenCourseWare*. Michael Perelman has observed that there are sound economic arguments for such an approach because, unlike traditional commodities, the unbridled spread of knowledge and information may well lead to the generation of yet more knowledge and information. In fact, this is one of the purported benefits of public education. Precisely where the intellectual commons begins and ends is not entirely clear. It has been argued that the creation and acquisition of knowledge is a kind of 'immaterial labour' that is fundamental to the fabric of our social life (Hardt and Negri 2000). Public education is a clearly part of the intellectual commons, but early language acquisition and the learning of so-called 'social skills' in the home might also be considered a part of this commons. The intellectual commons may even extend to the simple act of two academics having a chat about their work over coffee. This may seem to be going too far, but some scientists are already expressing concerns that it is precisely this kind of intellectual exchange that is being eroded by some of the new knowledge management practices (Marshall 1997). There are some fascinating legal initiatives being developed to extend and protect the intellectual commons as an alternative to existing copyright laws, notably the GNU General Public Licence (Moody 2001) and the Ordinary Publication Licence (*The Economist* 2 December 1999, www.opencontent.org). These forms of copyright allow for the creation of intellectual property for which no price may be charged. This copyright also gives limited property rights to the creator in cases where the original idea is being changed and developed in some respect. Similarly, the GNU General Public Licence has been developed for the distribution of freeware and is aimed at creating an intellectual commons on the Internet

(Moody 2001). These copyright initiatives are still a form of knowledge management, but they are designed to protect communal territories that are relatively unregulated with fewer barriers to the free exchange of knowledge and information. What is at stake in these struggles is the creation of an intellectual commons, which lies at the heart of the information economy. Where business is increasingly focused upon knowledge production, some view the creation of an intellectual commons as nothing less than a re-appropriation of the means of production (Hardt and Negri 2000: 303).

## The new cartographers of organization

The mapping metaphor is an important idea in diverse areas of study. In the management literature, the map has been used in the areas of operational research (Eden *et al.* 1979), corporate strategy (Weick 1994) and knowledge management (Colville *et al.* 1999, Nonaka and Takeuchi 1995, Davenport and Prusak 1998). This chapter has demonstrated that the metaphor of mapping is becoming increasingly important with respect to how we understand and manage knowledge and information. New practices and technologies associated with 'knowledge management' use mapping as a way of representing knowledge as a territory that can be defined, defended and regulated. This process of territorialization concerns itself with explicit knowledge or those aspects of knowledge that can be made explicit. Managers may be described as the cartographers of the organization, where maps can be used to make knowledge visible as a territory over which they can exercise sovereignty and control.

Knowledge management has been shown to privilege the visible and explicit aspects of knowledge, and tends to acknowledge tacit knowledge only insofar as it can be made explicit. However, a number of problems and points of resistance have been identified that are associated with attempts to control and territorialize knowledge. These points of resistance include the difficulty of isolating the creative process, the increasing alienation of product from producer and the economic contradictions of information markets, all of which lend inertia to attempts to create territorial maps of knowledge. Here, labyrinthine aspects of knowledge come into focus, since these tend to resist efforts of formal representation and commodification. These labyrinthine aspects tend to be portrayed in negative terms, simply as points of resistance that can be overcome given the right control strategy. Such strategies include Nonaka and Takeuchi's suggested use of dialectic and metaphor to convert tacit knowledge into explicit knowledge, or Davenport and Prusak's use of 'knowledge maps' to help locate sources of expertise within an organization, or Cook and Brown's 'generative dance', which takes place between tacit, explicit, group and individual knowledge to create new knowledge. However, the strategies proposed in this kind of work do not clearly address the four points of resistance explained above, notably the problems concerning creativity, alienation and the economic contradictions of information markets. This brings us to the final section of the chapter, which re-evaluated the role of mapping with respect to knowledge by drawing on the work of Deleuze and Guattari. This section provides a re-evaluation of the process of cartography, where the points of resistance are not seen in a negative light but are celebrated as a source of productivity. A distinction can be made between understanding the concept of a map as a territorial model of reality or as a labyrinth. In those areas of management thinking such as knowledge management and strategic management it is clear that the map is used as a model or representation, and the experimental and labyrinthine aspects of mapping have been marginalized. Alternative approaches to knowledge management are being developed which use the points of resistance as potentially productive forces, from which genuine alternatives for knowledge management are arising. For example, the post-structural concept of mapping as a labyrinthine activity that resists such territorial movements, where knowledge is not merely seen as a commodity but as an essential mediator in our self-understanding. Creativity concerns the posing of problems in the first place and not simply the problem solving process, which emerges secondarily. As part of a creative process, mapping can be seen as an experimental process rather than merely as a representational device. Some alternative approaches to knowledge management were then considered, which are aimed at creating and protecting an intellectual commons. These initiatives directly address the concerns regarding the fact that information is a non-rivalrous product that involves a number of intractable economic problems. These alternative approaches are not contrary to 'knowledge management' or symptomatic of a lack thereof, but are productive strategies in their own right. The proponents of these different approaches to knowledge management, from the mainstream map makers who are commodifying knowledge to those engaged in protecting an intellectual commons, are the new cartographers of organization.

# Part II

# The weapon of information

# 4 Organization, society and information warfare

War is the father of all things.

(Heraclitus)

. . . history progresses at the speed of its weapon systems.

(Virilio 1986: 68)

This chapter is devoted to the concept of information warfare. Information warfare is not only a new paradigm in military strategy, but comparable strategies are emerging in all kinds of civilian organizations. Commentators on information war, at its most simplistic level, believe that information and network forms of organization can significantly improve the capability of an organization to maintain its identity, to survive and prosper. These authors have attempted to develop an organizational model best suited to using the power of information and communication networks in both industrial and military contexts (Denning 1999, Toffler and Toffler 1993, Arquilla and Rondfeldt 1997).

The main aim of this chapter is to develop an adequate concept of information warfare that is appropriate to the kinds of strategies being employed in business and other civilian organizations as well as to the military context for which the concept was originally developed. Information warfare has been discussed in relation to several different fields, some choosing to focus on its military applications (e.g. Arquilla and Ronfeldt 1997), whereas others comment on its applications across the social spectrum (e.g. Denning 1999, Virilio 2000a, 2002). This chapter will draw heavily on the work from three main camps: (i) military analysts such as John Arquilla and David Ronfeldt, who produced the first major text on the subject, and Barbara and Alvin Toffler, who popularized the idea of 'information age warfare'; (ii) social critics such as Paul Virilio, who has written extensively on information warfare; and (iii) computer scientists such as Dorothy Denning whose work provides the most rigorous analysis of the concept to date. Much reference will also be made to the writings of the Zapatista leader, Subcomandante Marcos, who is without doubt one of the most effective practitioners of information warfare in the world today.

Information warfare is developing as a distinctive doctrine as a consequence of the revolution in information technology and the huge importance of information to the post-industrial economy. Information warfare refers to a diverse range of techniques, both high and low technology, which may be used to defend one's information resources or attack those of one's adversaries. These techniques include propaganda, public relations, intelligence, identity theft, and the disruption of movement and communication by virus or other means. One feature of these techniques that will recur throughout the chapter is the blurring of boundaries between the military and civilian, with the techniques of information warfare being developed and deployed in both domains. One of the chief arguments of this chapter is the central importance of identity in this new form of warfare. Information warfare operations, whether they are of military or civilian origin, take as their target the ideational, technological and informational structures of an organization. The main conclusion of this chapter is the need for individuals and organizations to become aware of the increasing threat of information warfare operations that may be pursued against them.

## Material conditions and the power of ideas

> An idea can only be fought with another idea or another concept.
>
> (Virilio 1999: 34)

To speak of 'the power of ideas' is important but only when considered in conjunction with the material conditions within which such ideas may thrive. The very possibility of knowledge is itself founded on a history of cultural practices. Nietzsche (1994), for example, explored how it was that the human animal developed a memory in the first place. He argued that various techniques of mnemonics and ascetic practices were used to pacify the human animal, and that this process must have been fearsome. Animal passions had to be domesticated and trained, and as a consequence the body became sickly in order to fashion it into a possible receptacle for knowledge. Some cultural practices may create environments that are better for the spread of ideas than others. Universities, for instance, appear to be ideal places for the spread of viruses, whether in the form of ideas, software, or medical conditions.

One should emphasize the fact that ideas are not powerful in themselves but require certain technological and material conditions for their spread or suppression. The ideals of the Enlightenment took root in soil far from the centre of colonial power in Europe. There was a degree of economic unrest resulting from the fact that the American colonies were subjected to the Stamp Act, which imposed the first direct taxation on that colony's produce. The Americas were also alive with dissident voices, such as the revolutionary protagonist Thomas Paine, who challenged the existing monarchic system and championed the founding of a free republic. These ideas could circulate within the colony in a relatively unrestricted fashion, and certainly played an important part in the struggle for American independence. The French Revolution took place in Europe shortly thereafter, during a period of poverty and repression. When food became scarce and the price of bread in Paris began to soar, the peasants began to attack food convoys on their way into the city. The peasants refused to pay taxes and tithes to their landlords, whom they held responsible for their economic plight. There was also a liberty of the press and the emergence of newspapers that printed popular ideas, such as the right to freedom of association and expression. Under conditions such as these, revolutionary ideals could flourish and spread rapidly and become the basis for the reorganization of society. The printing press was one of the earliest innovations in information technology, and shortly after appeared a prototype of information warfare – the revolutionary's pamphlet. In the eighteenth century, pamphlets were used to spur on the democratic revolutions of both America and France in spreading subversive and revolutionary ideals. The historian Barbara Tuchman (1984) attributes great powers to the role of pamphleteers and their ability to incite revolt under the right conditions. Perhaps the most famous of these pamphlets is Thomas Paine's *Common Sense*, which was one of the first written calls for American independence. According to Tuchman, this pamphlet 'had electrified the colonists, convinced thousands of the necessity of rebellion and brought them with their muskets to the recruiting centers' (Tuchman 1984: 263). Tens of thousands of such pamphlets were produced and disseminated in revolutions from America in 1776 to Russia 1917. They had an advantage over other media in that they were small and mobile and need not be tied to any particular publisher, and so were hard to shut down or suppress by the authorities.[1] Today, the website can be seen as a high technology version of the pamphlet, performing the same function as a medium for revolutionary and subversive ideas.

A materialist analysis of ideas can also be seen in the work of Marx and Engels (1998: 59), for example: 'When people speak of ideas that revolutionize society, they do but express the fact that within the old society, the elements of a new one have been created, and that the dissolution of the old ideas keeps even pace with the dissolution of the old conditions of existence.' One might add that if suitable material conditions are not apparent, then revolutionary ideas are bound to fail, which may well explain the apparent failure of Marxism in the former Soviet Union and other countries. Creating the conditions for the spread of information and ideas is a crucial feature of information warfare. During the slave revolt in the Caribbean, it has been remarked that the ideals of the Enlightenment travelled in a viral manner, from island to island, once the first outbreak had succeeded (Hardt and Negri 2000: 51). The virus was able to modulate its form when jumping from one island to the next, adapting to the specific conditions of each. Revolution became contagious. Henry Kissinger used exactly the same reasoning but for very different ends when excusing the human rights violations carried out in Latin America with the support of the US government. The democratic election of the Marxist president Salvador Allende in Chile was described by Kissinger as a virus that could infect other 'stable' countries in the surrounding area (Chomsky 1999). In order to deal with this localized infection, a CIA sponsored coup was effected in 1973 to replace the democratically elected president with a military dictator, Augusto Pinochet.

Despite Kissinger's best efforts, this virus has continued to flourish. During the 1990s in Mexico, a new rebellious force emerged from the Lacondon Jungle, comprising diverse indigenous groups in protest against a corrupt government and conditions of gross inequality and poverty that had been further exacerbated under the North American Free Trade Agreement. The leader of this rebel force, Subcomandante Marcos, has explained that the local mountains and jungle became an ideal ground for gathering those who were no longer willing to accept their miserable conditions, mainly the nameless and faceless peasants whose likely fate would be to die in abject poverty. This gathering began to forge a coherent identity under the name of the Zapatistas, harking back to the revolutionary leader who was instrumental in giving Mexico its constitution in 1917. These guerrillas have been remarkably successful in their use of the communications media, to such an extent that one guru of the information age, Manuel Castells, has described them as the first 'informational guerrillas' (Castells 1997). In his numerous writings, Subcomandante Marcos makes frequent reference to the power of words. He describes the task of their rebellion in metaphorical terms, where words and ideas are compared with seeds to be planted and people are compared with the earth in which the seeds may flourish. It is the task of the Zapatista rebellion to protect and cultivate this earth (Marcos 2001: 83–84). The Zapatistas recognize that those in positions of power in Mexico do not listen to the voices of the oppressed peasants, and by using the strong arm of the state – the police, the army and the juridical system – power actively silences these voices. The strategy of the Zapatistas, therefore, has hinged on attempts to keep broadcasting their message, 'When we were silent, we died. Without the word we did not exist' (Marcos 2001: 80).

The word can adapt itself and flourish in a diversity of environments. Many of the world's major religions originated in the barren wilderness of the desert, including Judaism, Christianity and Islam. The desert seems to provide conditions highly conducive for religious beliefs to flourish. T.E. Lawrence said of his experiences with desert peoples that, 'those who went into the desert long enough to forget its open spaces and its emptiness were inevitably thrust upon God as the only refuge and rhythm of being'

(Lawrence 1962: 39). He observed that the kind of faith that was fostered in the desert simply could not be found in the cities. Common to all prophets seems to be a period of abandonment in the harsh conditions of the desert. The parables of Jesus recognized the importance of material conditions for the dissemination of his ideas. In the parable of the sower, Jesus talked of the way the sower scatters seed and how some seed may fall on fertile ground suitable for growth, or it may fall on stony ground or become covered with weeds, both of which stunt the growth of the seed. This parable is a direct commentary on the spread of religious ideas, as Jesus himself explained, 'The sower soweth the word' (Mark, 4: 14). The life of Jesus makes a remarkable case study of a primitive yet subtle form of information warfare. Jesus preferred to speak in parables rather than to preach directly, an approach that facilitated the spread of his ideas, but the parables were also sufficiently ambiguous to mask his subversive message. He worked within the social and material conditions of his time to be most effective, and one need not be a believer to admire the subtlety of his strategy. In contrast to harsh desert life, the West has very different material conditions that are relatively comfortable for the majority of people who live there. Under such conditions, propaganda works primarily through entertainment and the news media (Herman and Chomsky 1988). Countries governed by radically different ideologies may go to extreme lengths to regulate their people's access to communications media. For example, the former Taliban regime in Afghanistan imposed strict prohibitions against the circulation of images, even destroying televisions when it came to power (Rashid 2001).[2]

There is a certain sense in which information warfare today has a democratic potential, as highlighted by the oft-cited example of the Zapatista rebels in Mexico (Castells 1997, Arquilla and Ronfeldt 1997, Taussig 1999). By stifling the flow of ideas and opinions, a government or organization may unnecessarily restrict internal variety, strangling innovation and damaging its capacity to adapt to changes in the wider environment. This has been put forward as a major reason for the collapse of the former Soviet Union, from commentators in both the mainstream (Toffler and Toffler 1997) and the radical left (Hardt and Negri 2000). The heavy-handed methods of social control exercised within the former Soviet Union lie in stark contrast to the relative free flow of ideas that encouraged the development of more advanced forms of information technology in the West.[3] The Tofflers summarize this problem in the following bizarre analogy, 'The Soviets, in fact, waged information warfare against their own people and shot themselves in the brain' (Toffler and Toffler 1997: xviii). The following analysis will therefore look at both the material conditions and conceptual elements that surround the practice of information warfare and the fundamental elements to its emerging form.

## Introducing information war

Information warfare is being taken very seriously by the military, where the current revolution in military affairs in the US is attracting millions of dollars of research funding (Arquilla and Ronfeldt 1997). The Russian military is taking information warfare so seriously that some of their analysts have reported that they would not rule out a nuclear counter-attack in response to an attack on their information infrastructure (Thomas 1996). The term 'information warfare' can be traced back to 1976 where it was used in the limited sense of 'degrading the enemy's weapon systems' (Arquilla and Rondfelt 1997b). It has been used in connection with diverse areas of study including, military strategy (Arquilla and Ronfeldt 1997), computing (Denning 1999), drugs (Plant 2000), business strategy (Adelstein 2001), social theory (Virilio 2000a, 2000b, 2002) and international relations (Der Derian 2002). This discussion will focus mainly on civilian organizations, but to accomplish this the argument will explore the concept of information warfare across the whole range of the previously mentioned areas of application.

The idea of information warfare has been the subject of a variety of definitions, where the only common denominator is the term 'information' and the fact that each definition reflects the background of the authors. One literature review of information warfare concluded with five different concepts being outlined where there was no clear resolution in favour of a single overarching definition (French 2000). Certain prejudices may be discerned in the existing attempts to define the concept, for example, the presence of a military flavour to many definitions, which is unsurprising given that this is the domain from which the idea first originated. Another prejudice is the tendency to highlight the high-tech forms of information warfare, which can be found in diverse writings including military analysts such as Arquilla and Ronfeldt, computer scientists such as Denning, and even social critics such as Virilio. Given the importance of high-tech communication and information systems to the economic infrastructure, it is not surprising that this has become a major strategic consideration for politicians and the military. Whilst recognizing its military origins and the implications of high technology for revolutionizing information warfare, we should not sidestep the possible use of low technology approaches, nor discount the use of similar weapons and strategies in the civilian world. Indeed, the Zapatista rebels in Mexico, who are seen by many as the pioneers of information warfare, originate in one of the poorest regions of Mexico from its most underprivileged and uneducated peoples.

A succinct definition of information warfare has been given by Arquilla and Ronfeldt at the beginning of their seminal work in this area: 'the use of information to impose one's will upon an adversary' (Arquilla and Ronfeldt 1997a: 14). This statement is interesting because it is as much an expression of the culture of Western individualism as it is a definition of information warfare. In certain respects, the above definition is itself an odd form of information warfare, based as it is on the exultation of the will. This highlights some of the difficulties that may be encountered in providing a comprehensive definition of information warfare, because such attempts tend to reflect a particular Western cultural milieu, which suffers from an individualistic ideology and technology fetishism. Although the focus of their work is on its military applications, Arquilla and Ronfeldt (1997a) note that information warfare concerns the technological, organizational, and ideational structures of a society. It is quite clear from such a statement that this new form of warfare is striking in its scale of conception, targeting the very fabric of society.

There is some agreement that the first information war was the Gulf War in the early 1990s, followed by the war in the Balkans, where Allied (US) forces refined this new strategy further (Virilio 2000a, 2002, Arquilla and Ronfeldt 1997). Certain technological and organizational innovations have allowed the development of the new doctrine of 'information warfare' by the strategic analysts of NATO and the Pentagon. The Tofflers point out that, in 1993, the first course in Information Warfare was initiated at the National Defense University at Fort McNair (Toffler and Toffler 1993: 164). West Point also appoints its own professors in Information Management and a recent copy of the US Army Field Manual (*Field Manual (FM) 100–5*) has stressed the changing nature of war and the crucial role of information. To quote from this army manual, 'Recent experiences gave us a glimpse of new methods of warfare ... They were the end of industrial-age warfare and the beginning of warfare in the information age' (Toffler and Toffler 1993: 62).

Technological developments are transforming the battlefield. During the first Gulf War, the air war alone required nearly 30 million phone calls (Toffler and Toffler 1993). The US coalition made use of around 60 satellites to conduct operations in the Persian Gulf (Toffler and Toffler 1993). Writing in 2000, Paul Virilio noted the ongoing presence of around 50 satellites in the space above the Balkans. He has identified three technological advances that have radically changed the conduct of war: the permanent presence of satellites, the real-time transmission of information, and the ability to perform a rapid analysis of this information (Virilio 2000a). This same technology has also played a major role in transforming the civilian world. The use of satellite technology and the rapid spread of computers has facilitated the globalization of communication networks and provided much of the infrastructure for the increased globalization of finance and commodity markets.

The RAND Corporation analysts Arquilla and Ronfeldt have coined a new term to describe information war in the civilian world: 'netwar'.[4] This kind of war tends to be conducted by relatively loose networks of people and organizations, often mediated by the Internet. Netwars are fought by small groups of people with shared ideas that can motivate action without the need for a central leader, and they are coordinated by means of electronic media, including radio, fax and the Internet. The example Arquilla and Ronfeldt use to illustrate their concept of netwar is the Zapatista uprising in Mexico in the mid 1990s. The Zapatistas were successful in raising awareness worldwide of the impoverished conditions of the peasant farmers in the Chiapas region of Mexico and in constraining the violent reaction of their government in its attempt to repress the protests. They achieved this by forming an alliance of diverse groups including peasant farmers, women's community groups, armed rebels, and academics, and also inviting in all kinds of NGOs from outside of Mexico, such as human rights groups. This complex network of alliances succeeded in getting the message of the oppressed indigenous peoples heard globally. Despite this worthy example, the RAND analysts are concerned that the same tactics are being employed by drug syndicates, organized crime and terror networks. They go as far as to say that 'The United States is fraught with divisive social netwars' (Ronfeldt and Martinez 1997: 374). So information warfare, specifically in terms of netwar, is as much about the internal governance of civilian conflicts as it is about conflicts with foreign powers. For the authors, it is clear that governments must develop new strategies to deal with this new situation and develop their own information warfare capability.

The most comprehensive formulation of information warfare in the civilian realm at present has been accomplished by Denning (1999), a professor in the field of computer science. She has outlined four main components to information warfare: information resources, players, offensive operations, and defensive operations. All four of these components revolve around the issue of security, which Denning places at the heart of her theory of information warfare. Information resources include containers, transporters, sensors, recorders and processors. All objects hold information, but the focus of information warfare tends to be on those objects that hold additional content, such as databases. Technology is currently revolutionizing the efficiency of all five of these resources, but it should be noted that human beings are exceptional in that they are adept at all five functions of these resources. The interconnection of all these resources and the flow of information between them is known as the 'information space'. Denning lists the 'players' of information warfare under six general classes: insiders, hackers, criminals, corporations, governments and terrorists. Each of these has their own distinctive motives for engaging in information warfare and has access to different kinds of resources. Business corporations possess massive resources to engage in information warfare, and Denning describes the information warfare practised by these organizations in the following terms:

> They engage in offensive information warfare when they actively seek intelligence about their competitors or steal their competitors' trade

> secrets through illegal means, such as bribing insiders. They sell information about their customers, sometimes violating their privacy. They are motivated by money and competitive position.
>
> (Denning 1999: 26–27)

The third component of information warfare is the list of defensive strategies available. Denning states that absolutely everyone practises defensive information warfare, from the individual who is concerned with the security of their bank details or their privacy, to more highly resourced organizations such as business corporations and government agencies. Defensive strategies involve the protection of information resources and can be carried out in a variety of ways, including prevention, deterrence, indications and warnings, detection, emergency preparedness and response. Denning emphasizes the crucial role of authorization and access in these defensive strategies, which she terms the 'CIA model'. The final component of information warfare concerns the use of offensive tactics. The aim of offensive information warfare is to damage or exploit an information resource 'with the objective of increasing its value to the offensive player and decreasing its value to the defensive player' (Denning 1999: 28). This may involve a range of actions including piracy, theft, perception management, identity theft, intelligence operations, sabotage, censorship, tampering and the fabrication of information.

Denning's work provides a comprehensive analysis of the various methods of defensive and offensive information warfare, with a particular focus on computer security and the costs pertaining to this problem. Indeed, one of the serious questions at the heart of this new form of warfare is the difficulty involved in measuring the cost and value of one's information resources (Denning 1999: 30). This makes it equally difficult to determine how much effort to expend in protecting these resources, or attacking others. A major benefit of her work is that it clearly shows that information warfare applies equally to military and civilian enterprises, particularly with respect to their increasing dependence upon, and use of, information technology.

In sum, it is difficult to arrive at a simple definition of information warfare. This is partly due to the fact that changes in the kind of practices that might make up information warfare are ongoing. Different authors have emphasized different aspects, for instance Denning has emphasized the technological aspects of information warfare, whereas Arquilla and Rondfelt emphasized its structural aspects, and Virilio its ideological and fetishistic aspects. Furthermore, it is clear from the above discussion that although military in origin, the concept of information warfare is rapidly gaining currency in the civilian world. The following discussion will therefore attempt a comprehensive analysis of the concept by breaking it down in terms of its associated practices using the three fundamental questions. (1) What is the object of attack or defence, is it information resources, or ideas and beliefs? (2) How is it done in terms of the different methods employed, such as the use of propaganda, intelligence, disruptive techniques, and associated technologies? (3) Who does it, whether it be military personnel, private citizens or business people? It is a discussion of the latter question to which we shall now turn: who are the agents of information warfare?

## Civilians: agents of information war

The concept of 'netwar' developed by Arquilla and Ronfeldt already highlights a blurring of the boundaries between the military and the civilian worlds. The book *War and Anti-War* also describes 'a civilianization of war and weapons' (Toffler and Toffler 1993: 216). There appears to be a civilianization of the military in terms of a perceived need for flexible networks rather than rigid hierarchies, and the use of civilian information technologies and communication networks. But there is also a parallel militarization of civilians. Civilians and the military increasingly share the same communications equipment, whether it be the Internet (originally a military application), satellite surveillance and communications, or mobile phone technology. As the Tofflers put it, 'for better or for worse ... the soldier and the civilian are informationally intertwined' (Toffler and Toffler 1993: 178). Civilians are, increasingly, not merely the subjects of information warfare such as propaganda, but are the very people who design these technologies and the operatives involved in pursuing information warfare.

The relationship between the military and the civilian has always been intimate, certainly since the origins of democratic society. Democracy, as it was originally organized in ancient Athens, was in fact a military state where its civilian leaders and military leaders were one and the same people, the *strategos*. In a comparable historical development, the installation of the French Republic after the people's revolution of 1789 was accompanied shortly afterwards by the creation of a conscripted citizens army in 1793. This was not an imperial enterprise but a revolutionary army fighting for the ideals of the republic, intent on banishing feudal oppression throughout Europe. It would, therefore, be quite presumptuous to claim that the two spheres of the civil and the military are clearly distinct.

The way we make war has always been implicitly related to the way we work (McLuhan 1964, Toffler and Toffler 1993). Early agricultural societies made war according to the means available to them and, owing to the severely limited logistics, armies tended to restrict themselves to seasonal campaigns. The object of ancient warfare also tended to be orientated around the destruction or possession of the adversary's agriculture. With the industrialization of production came the industrialization of warfare and its mechanization. Just as production became organized on a massive scale, so did destruction. Logistics, communications and weapons all tend to have similar processes of organization in civilian and military domains. Just as networks are becoming *de rigeur* in the civilian world, so are they being considered as more appropriate in the military world, specifically in relation to the increasing deployment of 'task forces' and 'special forces'. Both military and civilian organizations are exploiting greater flexibility on the one hand and an intensification of integration on the other. In 1993, Paul Strassmann was appointed Director of Defense Information in the Pentagon. What qualified him for this senior military post was his previous business experience as head of information services for the Xerox Corporation. This blurring of the boundaries between civilian and military affairs is also clear in the development of propaganda and in the intelligence services, both of which have been described as important elements of information warfare, as we shall see in the following sections.

## Propaganda and public relations

Propaganda is a major weapon of information warfare, and can be used by any country to exert a powerful influence on its civilian population. Mainstream commentators, such as Alvin and Barbara Toffler, have stated that propaganda was key in starting the war in Yugoslavia. They go so far as to assert that, 'Without the media, without television, the war in Yugoslavia would not have taken place. The triggering of the civil war was linked to the media, to the call to war by the media' (Toffler and Toffler 1993: 3). Local radio stations have often played a key role in whipping up unrest and inciting racial hatred in civil conflicts, such as the massacre in Rwanda in 1994 and the disintegration of the former Yugoslavia in the 1990s.

Governments take the issue of propaganda very seriously indeed and often take pains to control the words and images to which the public can have access, especially during times of conflict. In 1996, the US government founded the National Imagery and Mapping Agency with the initial intention of processing satellite images for the Pentagon and the CIA. However, only two years later the function of this agency and its 10,000 employees was broadened to monitoring the flows of commercial imagery as well, and it is now an obligatory passage point for all such satellite imagery (Virilio 2000a). The propaganda techniques used in the Gulf and the Balkans included the suppression of potentially subversive information at home, the distribution of subversive information to undermine adversaries, and the spread of contradictory data to paralyse the actions of both adversaries and protesters at home. The psychological operations branch of military intelligence is often involved in generating such propaganda. Dorril's history of MI6 quotes one agent admitting the role of their office during the first Gulf War in 'massaging public opinion into accepting controversial foreign policy decisions' (Dorril 2000: 766). Specifically, this involved propaganda exaggerating Saddam Hussein's chemical and biological weapons capability to dampen public outcry when the bombs started to fall. As events unfolded in the lead-up to a second war in the Gulf, it was interesting to observe the same games being played out in the media. During the Balkans conflict, the US deployed troops that were specially trained in psychological operations from Fort Bragg. These specialists in 'psy-ops' expand the traditional command and control function of military intelligence to include influencing the enemy's emotions, motives, objective reasoning and behaviour (Toffler and Toffler 1993: 165). In their analysis of post-modern war, the Tofflers actively propose the use of propaganda as part of their new 'peace form', with which they hope to do away with violent conflict. The Tofflers summarize propaganda in terms of six chief mechanisms: (1) the accusation of atrocities, (2) the use of hyperbole to describe what is at stake, such as 'freedom' and 'civilization', (3) the dehumanization and demonization of one's adversaries, (4) the polarization of perspectives, like the declaration, 'you're either with us or you're against us', (5) the claim to divine sanction,[5] and finally (6) the use of meta-propaganda to discredit the other side's propaganda. The Tofflers' analysis of propaganda is limited to military propaganda. For a more comprehensive discussion of the role of propaganda in society and its many techniques, the reader can refer to the extensive analyses of this topic by Jacques Ellul (1965) and Herman and Chomsky (1988). The argument of the present chapter will follow the work of Herman and Chomsky in showing that the techniques of propaganda are not simply confined to the execution of military operations but are a pervasive feature of the modern media system.

Almost all commentators on information warfare agree that propaganda is an important weapon in the arsenal of any organization (Arquilla and Ronfeldt 1997, Toffler and Toffler 1993). There is some nascent work on the role of information warfare within the business community (Denning 1999). These accounts discuss a range of issues, from protecting the integrity of the organization's technological information systems, to the use of propagandist methods to undermine competitors and the perception of their products. A recent article in the *Consulting Times* performed a relatively detailed examination of the use of information warfare and propaganda techniques in the speeches of Microsoft executives (Adelstein 2001). These speeches have been explained as part of a general campaign by Microsoft executives to undermine its competitors, most particularly Linux and IBM. The propagandist elements that were identified in these speeches include the manufacture of misinformation, the use of misleading and false evidence, and the attempt to exaggerate the risks of loyalty to the adversaries of Microsoft (Adelstein 2001).

The use of propaganda techniques within the business community is by no means new, and can be traced back to before the Second World War. Prior to the War the term 'propaganda' did not have the negative connotations that came to be associated with it after its use by the Nazis. In the 1920s and 1930s, the term was sometimes used in the social sciences and the public relations industry as a term of self description (Chomsky 1998: 181). The public relations guru, Edward Bernays, wrote books both on propaganda and public relations and confessed to changing the name of his work from 'propaganda' to 'public relations' because of the negative connotations that became attached to the former (Bernays 1933, 1952). Ironically, that very change was itself an act of public relations. Name changing is a simple and often used propaganda technique. After the nuclear accident at the Windscale plant in England, one of the reactions to dealing with this problem was to change its name to Sellafield (a far safer place of course). Naming and name changing is a powerful tool in the propaganda industry: think of the emergence of terms such as 'collateral damage' in the military or 'human resources' in the management literature.

The public relations industry has become massive. In the US alone there are an estimated 150,000 people employed in the PR industry, far outnumbering the actual number of journalists working in the same country (Perelman 1998: 17). In fact, journalists frequently turn to the PR industry for their information, and much of the news is manufactured directly in PR departments. Chomsky (1998) has noted that the influence of PR extends to our schools, sports leagues and universities in the form of sponsorship, advertising and sometimes ownership. Advertising and marketing can have comparable effects on attitudes. Trillions of dollars, almost 20% of business expenditure in the US, goes on marketing, i.e. changing people's attitudes and behaviour through the manipulation of information (Chomsky 1998). This may not really seem like a threat to democracy, but its effects on the media are astonishing. In 1992, a wide-ranging survey of 150 newspaper editors found that the vast majority of them regularly received interference over the content of their paper from advertisers (Perelman 1998: 17). This frequently led editors to change the specific content of a story, or in some cases to kill the story completely. In almost every case, this manipulation of the media concerned the publication of information that portrayed a corporation in a negative light or was in some way unfavourable to the corporate advertiser.

It is clear that public relations and propaganda have been active as a form of information warfare within the business community for many decades, and that it is a well-established practice. This discussion of PR and propaganda will now give a brief outline of the general form of propaganda derived from the work of Herman and Chomsky (1988). They identified five 'filters' which they found to have an influence on the content of the media in their study of the US system. These filters are: ownership of media, advertising, sourcing, flak and anti-communist ideology. Chomsky and Herman's work found much evidence of the first filter in operation, and most notably there has been an increasing concentration of ownership of the media in the hands of a few powerful corporations, ownership being the simplest and most direct method of control. There is also a great deal of evidence to highlight the influence of advertisers on the content of the media, since much of a newspaper's revenue comes from this indispensable source. In fact, many newspapers actually sell more advertising space than they set aside for news stories. The third filter, sourcing, concerns the use of a limited range of experts and official sources of information, which are presumed to be accurate, but there is no attempt to verify the official version of events or listen to contrary voices. The fourth filter is flak, whereby dissenting voices are discredited or suppressed. The filters of sourcing and flak are also tied in to the huge influence of the public relations industry and the large number of stories that issue directly from such sources. The fifth filter of anti-communism has exerted a powerful influence in the Western media, although most intensely in the United States. However, in a recent article Herman (1998) has acknowledged that the anti-communist filter is likely to have diminished somewhat with the fall of the Iron Curtain. Having said that, my Microsoft word processor has picked up the word 'anticapitalist' as a spelling error but not 'anticommunist'. Should we conclude that this software is reflecting the prejudices of its programmers, or should we simply say that it is now out of date? Each of the five filters can have a small effect when considered in isolation, but collectively they are termed the 'propaganda model'. The whole idea of this model is that it acts as a network of constituent relations, governed by the market system. That is to say, the propaganda model proposed by Herman and Chomsky is not centrally controlled either by the government or by any other private corporation. Nonetheless, the market can have the effect of systematically distorting the way information is presented, by the action of one or more of the filters discussed above.[6] Naomi Klein's recent best-seller *No Logo* is crammed full of similar examples of how the market system, through the action of large corporations – such as Wal-Mart – is creating a climate of what she terms corporate censorship and linguistic and cultural privatization (Klein 2000: 177). This censorship is not confined to the work of PR departments, but is also encouraged by fear of offending one's employers or customers and by means of the aggressive use of copyright and trademarks against those who attempt to criticize certain powerful brands. Now that we have explored the role of propaganda and public relations as a powerful technique in the armoury of information warfare, we will take a brief look at the associated area of intelligence and its changing role in the business world over recent decades.

## Intelligence and secrecy

Secrecy is becoming the litmus test for the value of information. This has always been the case for military organizations, for instance, the US military has a strictly defined hierarchy of access: For Official Use Only, Confidential, Secret, NATO Secret, Top Secret, NATO Top Secret

(Toffler and Toffler 1993). It has been estimated that the United States government spends an impressive $16 billion dollars per year and employs 31,000 full-time workers just to safeguard classified documents (Perelman 1998: 77). One of the chief problems for organizations in the information age is how to measure the value of information. Consulting services and reports are becoming increasingly highly priced and there appears to be a clear relationship between the sensitivity of information and its accessibility.

The idea of 'intelligence' is becoming extended to the whole of society, particularly with the advent of netwars and cyberwars and the blurring of boundaries between military and civilian. Industry is gradually becoming infiltrated with ideas and people from the intelligence world. Stephen Dorril's (2000) history of British military intelligence describes situations in which intelligence agents have been used to win lucrative overseas contracts. In the UK during the 1990s, the Intelligence Services Act 1994 revealed that MI6 was involved in protecting the economic interests of the UK, particularly with respect to strategic resources such as oil and metals. Dorril's history of MI6 explains that this organization gathers useful commercial intelligence, which is passed on to Britain's major companies, including City banks, defence exporters such as British Aerospace, the oil companies BP and Shell, and other global companies such as British Airways (Dorril 2000: 761). Dorril discovered that in the late 1990s MI6 headquarters had been running an operation called 'JETSTREAM' which directed economic espionage against various countries, including France, Germany, Spain, Italy and Switzerland. (Dorril 2000: 778–779). Dorril notes that this has led to international tensions, and mentions the German government's annoyance with the activities of MI6, and particularly the CIA who were engaged in economic espionage within Germany's borders in the late 1990s. In the USA, the FBI has formed its own unit, called the 'Economic Counterintelligence Program', to deal with economic espionage. This unit has witnessed a huge leap in illegal corporate intelligence operations over the past few years (Denning 1999: 64).

After leaving military intelligence, former agents often transfer their skills into the world of big business working for oil companies, banks and the like. There is evidence of a 'privatization of intelligence' since the collapse of the Soviet Union (Sigurdson and Tagerud 1992). Military Intelligence has long been interested in economic matters that are considered essential to the security of the nation state, such as the arms industry, oil supplies, high-tech research and intellectual property rights (Sigurdson 1992). Even economic trends and the nation's agriculture may be seen as important components of a nation's security (Colby 1992: 20). The communications infrastructure of a nation is itself of strategic import. Hence recent legislation to protect the integrity of the nation's communication infrastructure, especially with respect to the privatization of many previously nationalized telecommunications monopolies. Security issues are a key part of the development of telecommunications legislation, for instance in the US the National Security Telecommunications Advisory Committee was formed in 1982 followed by The Telecommunications Act passed in 1996 to deal with deregulation, and in the European Union the European Cybercrime Forum was recently established over and above the legislation of individual countries, such as the UK's Regulation of Investigatory Powers Act 2000.

Corporate intelligence, espionage and counter-espionage are proving to be an important part of information warfare in the business world (Denning 1999). There have been arguments for the increasing importance of intelligence to business activities, where some have gone so far as to claim that 'war is not very different from fair competition' (Piganiol 1992: 34). The Tofflers have stated that the First World is already waging a form of information warfare against the Third World by poaching its intellectual resources (Toffler and Toffler 1993: 168). They cite the management guru, Tom Peters, in this respect who positively encourages Western business to poach 'human capital' from Third World countries. Another important form of poaching is the recent emergence of 'bioprospecting' where Western pharmaceutical companies are patenting genetic materials, both plant and animal life, which occur naturally in the ecology of Third World countries (Ostergard *et al.* 2001). Back in the early 1960s, Marshall McLuhan explained that tensions surrounded the development of new technologies, because both new weapons and tools may be perceived as a threat to all those who lack them (McLuhan 1964). This tension is only too apparent with the dawn of the information age and information warfare.

## Weapons and tools

> Weapons and tools are consequences, nothing but consequences.
>
> (Deleuze and Guattari 1988: 398)

Information technologies can be used as both weapons and tools, and they play an increasingly important role in industrial and military systems. The division between weapon and tool is, to a large extent, determined by the actual use of the object rather than being in the nature of the object itself. A simple illustration is that of a knife, which can be used as a tool to slice bread or as a weapon with which to stab someone. The difference between these uses can be broadly categorized as being either projective, in the case of a weapon, or introjective in the case of a tool (Deleuze and Guattari 1988). There is a sense in which all weapons are projective, which is clear in the case of weapons such as the arrow or spear. Projection is also implicit in the weapon that does not leave one's hands. For example, the sword cuts outwards, projected from the wielder's centre of gravity while sweeping around in an arcing motion; it is projected against the vital force of an opponent but it never leaves the hands of its wielder.

In the context of work, the tool is essential to the organization of the productive forces of labour and the disciplining of these forces. Labour power is assembled around the use of tools within a work organization. By means of such discipline surplus value can be extracted from the productive efforts of labour. The weapon, on the other hand, is used in an extensive manner to disrupt the organizing forces of the adversary. Both the weapon and the tool require discipline in their usage; however, surplus value is not contained or extracted from the use of the weapon. The use of weapons in the martial arts is not only about learning to do things, but also learning to undo things (Deleuze and Guattari 1988). In particular, the martial arts provide a way to undo oneself, or as the *Book of the Samurai* puts it, 'I do not know the way to defeat others, but the way to defeat myself' (Tsunetomo 2000). In contrast, there is something very domestic about the use of tools which is not present in the use of weapons. The weapon is organized on what Deleuze and Guattari term the 'free action model'. The weapon is essentially mobile, where the wielder has abstracted the idea of free movement from the hunted animal, and uses it as the motor force driving the weapon (Deleuze and Guattari 1988: 396).

This chapter began by pointing out that information warfare is associated with a blurring of boundaries. It is certainly true that today, with the increasing focus on the defence of one's information resources and the direction of attacks against the resources of others, weapons are becoming increasingly integrated into the work model of organization. Information systems are both tools and weapons, but in the case of information warfare, the system is no longer confined to the organization of productive forces, but is released onto the world for destructive and disruptive purposes. As a weapon, information systems are a kind of prosthetic motor force, a new way of moving, and projecting, which is at once viral, virtual, communicative, and terminal.

## The technology of information warfare

Much of the existing work on information warfare tends to focus primarily on high technology surveillance and communication systems, the Internet and the mass media. However, the armoury of information warfare has a diverse range of both high and low technology weapons that can be used to inform, misinform, manipulate perceptions and disrupt communications systems. Pamphlets and websites may be used for the publication of alternative ideas and viewpoints. As previously discussed, the mass media is still a powerful forum for public relations and propaganda exercises. Denning (1999) lists a number of Internet-based forms of information warfare including hacking, the spread of viruses, email bombs and virtual sit-ins. By the use of email bombs and virtual sit-ins one can bring down targeted information systems by bombarding them with messages and website hits, and overloading their capacity. By means of hacking one can gain access to vital information stored on an adversary's computers or sabotage the same system. Hacking has become a huge problem for diverse institutions including government agencies, the military, business corporations, banks and other financial institutions. Hacking often goes undetected, and computer systems may be hacked without the owner ever knowing their system has been compromised. Hackers often use false or temporary identities to avoid being tracked. Identity theft is another common form of information warfare, which works by stealing someone else's identity details or access codes, such as their name, social security number, credit card and bank account numbers. One of the biggest problems of such forms of information warfare is that they can originate from any part of the network, which has now grown to massive proportions. In the UK, the police are becoming concerned at the increasing level of Internet fraud originating in foreign countries, for example the *Financial Times* newspaper has picked out Nigeria as a breeding ground for Internet fraud and other kinds of information warfare (*Financial Times* 3 March 2003). Although it is not entirely clear why the FT highlighted this particular country as an exemplar of banking fraud, it is doubtless true that attacks on British banks can originate from criminals working in other countries, which have different, possibly more relaxed financial regulatory systems. A variety of means are being used to perpetrate these frauds, and sometimes the information is sent from people working within the target institution. Another method involves the setting up of websites that appear to be from reputable banks, such as www.barclaysprivate.com, or www.eurocibank.com, but are in reality bogus sites designed to steal the credit details of unsuspecting customers. Policing these problems is further complicated because financial institutions are reluctant to admit to frauds that might damage their reputations and undermine customer faith in their business. Instead, losses tend to get covered up as 'bad loans'. Legislation to protect against such attacks varies from country to country, which makes regulation of this media extremely difficult.

However, not all information warfare relies on the fruits of the recent technological revolution. A focus on high-technology methods, especially by the information rich, can sometimes lead to absurd applications in practice. Take for instance a story that featured in the *New York Times* from 24 February 2003 on the use of information warfare in the recent military campaign in Afghanistan. The use of propaganda against target populations is standard operating procedure to undermine morale and encourage desertion and surrender. In the case of Afghanistan this turned out to be a serious problem because the indigenous population did not know of the existence of the Twin Towers, let alone that they had been destroyed; so they did not understand why they were being bombed. In response to this the US military set up radio stations in Afghanistan to broadcast their message and explain the reasons behind their invasion. Alas, they had not counted on the fact that it was such a poor country that few people had access even to a radio. The next step taken by the military was to drop 500 radios on the unsuspecting population, all of which broke on impact. Although information warfare is not an absurdity in itself, this example does illustrate the obvious limitations that derive from too great a focus on high technology. The greatest prophet of the information age, Marshall McLuhan observed that the poorer nations might, in fact, hold some competitive advantages over their richer neighbours,

> In the new electric age of information, the backward countries enjoy some specific advantages over the highly literate and industrialized cultures. For backward countries have the habit and understanding of oral propaganda and persuasion that was eroded in industrial societies long ago.
>
> (McLuhan 1964: 366)

This has serious implications for information warfare, especially with respect to low-technology information warfare. McLuhan's observation partially explains the success enjoyed by the Zapatista rebels in Mexico, who have earned the title of the first informational guerrillas (Castells 1997). These rebels have used the oral tradition of the local Mayan culture to fashion an identity mixing history and myth to form their *Zapatismo* ideology. The name of Emiliano Zapata who died fighting against the powerful Mexican land owners early in the twentieth century, and whose example the Zapatistas follow, is also called 'Votan Zapata' and 'Ik'al Zapata'. A name from the twentieth century is combined with the names of ancient Mayan gods, one who lives by day and the other by night. This blend of myth and history is an assertion of the local indigenous culture but it also highlights the continuity of a struggle that has been going on since the invasion from European colonialists over five centuries ago. Incidentally, the metaphors of light and darkness are often used in Zapatista writings to show that *Zapatismo* is for people regardless of their skin colour and is by no means exclusive to the indigenous Mexicans. Their low-tech rebellion incorporates all aspects of their culture and their itinerant existence in the jungle and mountains of Chiapas:[7]

> ...we have the arm of the word. We also have the weapon of our culture, of being what we are. We have the weapon of music, the weapon of dance. We have the weapon of the mountains that old friend and campanera who fights along with us, with her roads, hiding places, and hillsides, with her trees, with her rains, with her suns, with her dawn, with her moons...
>
> (Marcos 2001: 168)

The distribution and use of high- and low-technology weaponry is not simply arbitrary. One of the characteristics of the information-age has been a widening access to certain forms of information, especially since the advent of the Internet, but even this has been largely restricted to the more middle class members of Western society. The most advanced high-technology weapons of information warfare, such as expensive surveillance equipment, are available only to the most powerful strata of society, and it is often the poorer strata of society who are the objects of this warfare. Low-tech weaponry is often taken for granted, perhaps because it is such a common feature of everyday life. In the intelligence community this is called 'HUMINT', or human intelligence, and it requires the use of front line agents (often journalists) with their ear to the ground to pick up on potentially useful gossip and rumour, and perhaps, to infiltrate target organizations. The writer William S. Burroughs has proposed a type of information warfare that makes much use of low-tech weaponry in a kind of bargain basement approach. This includes the use of methods such as the 'cut-up' and the spread of rumours to subvert the messages disseminated by the mass media. Indeed, almost all of us will be familiar with the power of rumour, a weapon that can be used to devastating effect. An extraordinary case of the influence of rumours is that of Jonathon Lebed, a 15 year old, who in 1999 used the Internet to spread rumours in an attempt to promote the value of penny shares he had brought. In a six-month period he had made a profit of around $800,000 and was subsequently brought to the attention of the Securities and Exchange Commission (SEC). This case has also caused the SEC to reflect on the very awkward problem of how to determine whether a rumour is legitimate or whether it should be classified as illegal manipulation of the market.

The global circulation of information and wealth is restricted by means of access codes and passwords. But one of the most notorious hackers of the early days of the Internet, Kevin Mitnick, has shown that a person can gain access to valuable corporate information without ever having to touch a computer. Mitnick (2002) tells us that the weakest link in most information systems is the human being rather than the technology. Hackers can use a mixture of technological know-how and social skills when gaining access to a restricted system. Mitnick uses the term 'social engineering' to distinguish the use of psychological manipulation from the purely technical skills of the hacker. The confidence trick is the social equivalent to hacking, whereby the hacker gains access to information under false pretences. Mitnick's work describes a series of cons that individuals and business organizations are frequently subjected to, all of which work by means of deception and psychological manipulation. Common tricks he lists include, among many others: posing as an authority, name-dropping, using the corporate jargon and local knowledge, and posing as a new employee. Often a hacker will use a telephone call to engage a victim and extract information, preferring to avoid direct physical contact. Mitnick also provides a list of symptoms to look out for which may indicate that some kind of social engineering is taking place, including the refusal of a caller to leave a callback number, an out-of-ordinary request, a stress of urgency, name dropping, a claim of authority, and flirting. The problem with this is that the symptoms he lists are hard to tell apart from everyday social interaction.

Denning's (1999) analysis of information warfare notes that whilst the delivery mechanism may be high-tech, the method itself may be fairly low-tech. Denning does not examine low-technology forms of information warfare in any detail but she does point out that: 'Practically any speech act affects perceptions, so the possibilities for information warfare are endless' (Denning 1999: 102). Speech acts may be seen as information warfare because they can be used to affect the perceptions of target groups, as apparent in propaganda, public relations and 'psy-ops'. In the highly individualistic terms of Arquilla and Ronfeldt, information becomes a weapon when used to impose one's will upon an adversary (Arquilla and Rondfelt 1997). More precisely, ideas become weapons when they are used to fight other ideas. In her extensive analysis of information warfare, Denning has expressed great concern over the unbounded nature of this form of warfare.

Speech acts are linguistic declarations that enact some form of social commitment or obligation. For example, a promise enacts a social obligation as it is uttered. The domain of speech acts is vast, including commands, questions, promises, threats, accusations, pleas, and any other words that bring with them implicit social presuppositions. These presuppositions are often characterized by asymmetrical power relations built into the social fabric, such as those between a parent and child, or between doctor and patient, or judge and criminal. The performance of interrogation is bound up in power relations where the inquisitor stands over the speaker or confessor (Foucault 1981). The inquisitor, whether they are a judge, doctor, or psychoanalyst, demands the truth from the confessor and at the same time acts as the interpreter who determines the truth of the speaker's statements. The power relations that underlie these speech acts are most obvious when someone who is not in a position of authority attempts to draw on this authority. Such situations are considered absurd or laughable, and might even be punished, such as the 'cheeky' child who answers back, or the patient who tapes the questions of his analyst, or the accused who is found in contempt of court. In his book, *Crowds and Power*, the Nobel laureate, Elias Canetti was very much concerned with the force that lies behind the act of commanding. He observed that commands are older than language itself, since even animals (dogs for instance) can understand commands despite having no linguistic capacity. Canetti described the effect of the command as 'the sting of the command' because we feel its effects on us – its sting – even after obeying. The sting of the command is left behind in us because it hits us as an alien force whose intrusion reminds us of our own subservience.

Deleuze and Guattari (1988) show that language is teeming with speech acts that operate within and bring to bear certain power relations. Obligations do not arise solely from commandments or condemnations but occur in all sorts of speech acts – 'what do you think you are doing?' (stop doing it!), 'really?' (tell the truth!), 'read my lips' (I'm not lying!), 'you are a man now, not a child' (stop playing!), 'there's no room for politics in this organization' (obey!).

Passwords are terms that carry an obligation with them, knowledge of which can give one access to restricted information and wealth. Deleuze and Guattari (1988) have defined the effect of such speech acts in terms of creating an, 'incorporeal transformation'. For example, when one says the words 'I do' at a wedding, one is transformed from being a single person into a married person, with a new set of rights, obligations and tax breaks. When sentence is pronounced upon an accused by the presiding judge in court, the accused is immediately transformed into a criminal, a change to their very identity. The speech act brings to bear a certain set of institutional power relations, like when the teacher tells us to be quiet or when the politician labels a group as terrorists rather than freedom fighters. Language does not merely describe social reality, but plays an interventionist role in terms of labelling and transforming that reality. The meaning and power of a word is largely determined by its non-discursive presuppositions, which are a part of the institutional apparatus in which the word is spoken. If an employee were to give a motivational speech to their manager on the office floor, or if an accused were to pronounce sentence upon the judge in the courtroom, these utterances would sound ridiculous. Deleuze and Guattari (1988) have argued that order words are not merely a subset of language, and that some obligation or non-discursive presupposition is implicit in all language.

Slogans and propaganda are clear cases where language plays an interventionist role. It would be absurd to suggest that all speech acts are also acts of information warfare; however, speech acts are clearly an important element of information warfare. When individuals or organizations use speech acts to further their aims, as in the case of slogans and propaganda, this is part of a campaign of information warfare. As has already been pointed out by Denning (1999), these speech acts are specifically aimed at changing our perceptions. The founding father of public relations, Edward Bernays (1923), stated that the proper function of public relations was to turn language into a tool of social control, where ideally the target audience would respond in an automatic stimulus-response fashion to certain words and images. The task of the PR man was to create and manipulate social stereotypes that could be used to short-circuit conscious reflection in precisely this way. Perhaps the greatest works about the interventionist powers of language are George Orwell's *Nineteen Eighty-Four* in which he outlined his distopian concept of 'newspeak', Ellul's analysis of the role of propaganda in society, and, more recently, Subcomandante Marcos' writings on the struggles of the Zapatistas (Ellul 1965, Marcos 2001).

The Zapatista rebels in Mexico have relied greatly upon the power of speech acts and slogans proclaiming that, 'our word is our weapon' (Marcos 2001). They have used a number of slogans as part of their ideological campaign, some harking back to the ideals of American Independence and the French Revolution, such as 'Liberty, Democracy, Justice!' – a slogan that is repeated hundreds of times throughout the writings of Marcos (2001). Other slogans have been tailored to the peculiar nature of *Zapatismo*, such as 'Enough is enough!' and 'Everything for everyone, nothing for us!' (Marcos 2001: 20). Slogans can act as powerful speech acts that mobilize action. The slogan 'Workers of the world unite!' has reverberated around the world ever since its first publication in 1848 (Marx and Engels 1998). Political parties of any persuasion recognize the importance of these kinds of speech acts, as indeed, do the advertising agencies whose services they increasingly seem to require. Following the simplest of Freudian techniques, advertisers attempt to exploit lines of association which they try to assemble in our unconscious. During the dot.com bubble of the late 1990s, which led to a massive inflation of shares in high-technology firms on the NASDAQ, a host of slogans began to emerge to justify this largely illusory era of unlimited growth. These slogans referred to the recent technological innovations, like the 'information super highway', but went marching into a brave new world and a new financial bubble with slogans such as 'e-solutions', 'the surge economy', and being 'the dot in .com' (de Cock *et al.* 2001). When faith in such slogans and their relation to the market began to waver, the bubble burst leading to a dramatic fall in the value of the NASDAQ from around $6000 billion to $3000 billion within a year. Slogans form an important part of our social fabric, and can help serve to unite interests and motivate action. However, as the example of the collapse of the NASDAQ highlights, they can mislead just as well as they can motivate. Lenin's 1917 text on slogans fully appreciated the power of words, but paid specific attention to their limitations within the material context. In his discussion of the proletarian struggle for democracy he was concerned about 'slogans which had formerly been correct but had now lost all meaning...' (Lenin 1964: 183). The authorities reacted to the demonstrations and public protests of the masses not by using arguments but by turning to armed suppression. In the face of such violence, Lenin argued that it was not enough for the masses to protest peacefully for power and democracy or to repeat slogans to this effect, because they needed to rise up and physically seize power. Mere slogans were not enough, which became clear when those who had been chanting them were gunned down by pro-monarchist cadets. Lenin's lesson was that whilst they are a powerful motivating force, out of context slogans could be mocking, or worse still, deceiving.

Information warfare has emerged as a distinctive concept during a period of massive technological upheaval in the means of communication and information technology. This section has shown that information warfare is not simply a spin-off of high-technology industries, but exists as a range of both high- and low-technology weapons, employing simple techniques such as rumours, slogans and speech acts, but amplifying the effects of such weapons with the high-technology of computers, the Internet and the mass media. The discussion of low-technology information warfare also highlights the fact that it directly concerns our language, culture and self-understanding. What distinguishes information warfare as a concept is not any single technique, but the fact that we have crossed a technological threshold, where many of the targets of defence and attack in both civilian and military organizations are information resources. Information resources and technology have become the objects of attack and defence, but are also the primary means for carrying out such attacks.

## Speed and information

> We have developed speed, but we have shut ourselves in.
>
> (Charlie Chaplin, *The Great Dictator*)

> Speed is the essence of war.
>
> (Sun Tzu, *The Art of War*)

Speed has societal implications both in terms of a nation's relationship with external powers, and in terms of its internal relations – the organization of society itself. It was the strategic genius Napoleon Bonaparte who observed that 'Aptitude for war is aptitude for movement' (Virilio 1986: 22). The ability to mobilize an army and carry out manoeuvres is crucial to military success in any war. One of the tragic aspects of the Great War was the paralysis forced onto all sides. Warfare has always been concerned with movement and the disruption of movement. Certainly, the new capability afforded to the military forces of the West by modern information systems has transformed their ability to manoeuvre. Speed plays a key role in this respect. Modern technology speeds up the way in which warfare is conducted in terms of the capacity to deploy troops, vehicles and weapon systems and the speed of intelligence gathering and communication. Speed makes war convenient (Virilio 1986: 55).

The earth itself has become less of an obstacle, both in military and industrial terms, with the twentieth-century revolution in transportation and logistics. The rapid transfer of goods and raw materials to the market has been achieved on an international scale by means of seaports, airports, road and railway. With speed, the idea of physical territory has diminished in importance. Speed is now directly concerned with the control of time rather than space, particularly with respect to its recent manifestation in technologies of 'real-time' control. In his exegesis on speed, Paul Virilio describes speed as a 'non-place', which opens up novel questions concerning the nature of territorial appropriation (Virilio 1986: 133). In many respects Virilio's commentary, which was first published in 1977, anticipated the emergence of the Internet a decade or so later. This non-place of speed of which Virilio wrote has some resonance with the virtual reality provided by the Internet, where people come together to communicate and exchange ideas on a new global level.

Speed has been an important part of the transformation to an information based economy, where industry looks to speed in order to remain competitive (de Cock *et al.* 2001). Speed is crucial to the information economy and is an important determinant of the value of information. Gaining information in advance of one's adversaries is an important strategic consideration in both military and business situations. The timeliness of information is a major component of its value; it is precisely speed that guarantees both the secret and thus the value of information (Virilio 1995). Commentators on the information economy now claim that economies of speed are as important as economies of scale (Toffler and Toffler 1993: 72). Small-scale operations may not have the resources of large organizations, but with relatively cheap information technology they can still achieve the same or greater economies of speed. In the modern management literature speed is central to practices such as Just in Time management, Supply Chain management and Enterprise Resource Planning. These management practices use information technology to enhance real-time remote control to further integrate standard operating procedures with computerized control systems. Intensified interactivity minimizes the time lapse between the command and its execution, such as between the point of sale and the stock re-order, or between the identification of a target and its destruction. Paul Virilio is concerned that this increased interactivity is removing the time for reflection, which he believes to be necessary for human freedom. He warns us that, 'the more speed increases, the faster freedom decreases' (Virilio 1986: 142). In this respect, he proposes personal reflection and face-to-face interactivity as ways to slow us down and give us pause in order to provide time for freedom.

Although Virilio tends to emphasize the distopian elements of a society obsessed with speed, he does provide an alternative concept of speed in his analysis of guerrilla warfare. Guerrilla warfare takes the control of time as its object rather than the control of space (Virilio 1990: 50). This is how the Spanish guerrillas fighting Napoleon's army gained their advantage, when they did not have the advantage of controlling their own country. Guerrillas simply do not have the power to fight occupying forces by defending a fixed enclosed space. Instead, they master time; they can choose when they will fight, and they have the advantage of untimely surprise, speed of attack and speed of dispersal. By means of speed, they create a space within which to move unimpeded. This was often the case in past wars where guerrilla forces could evaporate into the desert without leaving a trace (The Arab Revolt of the First World War), or melt into the jungle and hidden underground tunnels (Vietnam), or hide in the houses of civilians in the case of urban warfare (Northern Ireland and Palestine). Deleuze and Guattari (1988) have distinguished between three forms of speed: (i) nomadic or revolutionary speed, such as that of the riot or guerrilla warfare; (ii) regulated speed, such as found on the public roads that are regulated by state agencies; and finally, (iii) the speeds that result from total war, and movements which encompass and threaten the entire globe. As Sun Tzu tells us, speed is a defining characteristic of warfare, it is an element that is essential for the purposes of control and the disruption of control. It is therefore clear why the new communication technologies are playing an increasing role in both industry and warfare in their pursuit of ever-increasing speeds.

## Disruption and the internalization of information warfare

One of the key components of the doctrine of information warfare is the fact that the requirement to destroy recedes as the ability to disrupt is enhanced (Arquilla and Ronfeldt 1997a). This may give the most powerful countries the opportunity to be relatively merciful in conflict. Having vast information superiority, some countries, such as the US, can sit back and engineer victory as much as possible by the remote guidance of confusion (Virilio 2000a: 49). Virilio has observed the era of deterrence is drawing to a close but with it dawns a new era of information war, where the superiority of information is more important than the ability to inflict damage (Virilio 1997a: 175). At the same time, smaller enemies may also be empowered by the ability to disrupt, which has always been an important component of guerrilla tactics. New opportunities for such resistance are emerging where 'the more technologically advanced an opponent is, the more he may be vulnerable to disruptive attack' (Arquilla and Ronfeldt 1997a: 13).

By disrupting enemy movements, one attempts to multiply the number of accidents they experience in their everyday activities. Virilio (1999) has observed that the accident is the hidden face of technical and scientific progress. Early in the industrial revolution Marx recounted the horror of a rail crash that 'dispatched hundreds of passengers into the next world' (Marx 1976: 363). The same high speeds that are revolutionizing capitalism are also revolutionizing the kind of accidents and disasters to which we are being subjected. The effects of disasters can become amplified on a massive scale due to the networks themselves. Take, for instance, examples such as the now outdated Y2K computer bug, pandemic outbreaks of viral contamination (flu and AIDS), and the recent concerns that have been voiced over the uncontrolled contamination of natural plants and animals with genetically modified products (Quist and Chapela 2001). Global communication networks are facilitating the spread of different viruses whether they are virtual or biological in nature. Financial bubbles are yet another example and have become a recurrent feature of the world system, exemplified by international stock market crashes, from the South Sea Bubble of 1720, to the crash of the Asian Tiger economies in 1997, and the burst of the 'dot.com' bubble of 2001. All of these examples highlight the increasing danger of a new kind of disaster, distinctive because it will occur everywhere at the same time (Virilio 1999: 92).[8] Virilio has an apocalyptic view of the future in which what he terms the genetic and information bombs are now forming a single weapons system that threatens the constitution of life itself (Virilio 2000b: 140). Virilio's view of the future may be overly pessimistic, but it is nevertheless true that the principle of disruption that characterizes information warfare is becoming internalized as a part of our social fabric, and is an everyday feature of global communication networks, specifically in terms of software bugs and viruses. The doctrine of information warfare specifies that disruption is one of the key methods by which one can attain one's goals, specifically by intentionally disrupting the information resources of one's adversaries and so confusing their decisions and movements. The flip side of this, however, is that in a networked world, the effect of local disruptions and unintended accidents may also proliferate.

## Violence, terror and the omnicrisis

> Fear is the cruelest of assassins: it never kills, but keeps you from living.
>
> (Virilio 1986: 39)

Both the military and civilian dimensions of information warfare are comparable in their focus on the disruption of communication and movement. Some have claimed that the new approach to conflict offered by information warfare will lead to a reduction in the violence of the wars of the future (Toffler and Toffler 1993). There are at least three reasons given to support such a view. The first is that the direct use of propaganda can diffuse dissent at home and violent resistance in foreign territories. The second reason is that the pinpoint accuracy in weapons guidance enabled by the new information technologies should lead to significant reductions in so-called 'collateral damage'. And the third reason is that there has been a proliferation of non-lethal technologies, which can be used to disrupt the movements of one's adversaries. These non-lethal weapons include the use of computer viruses to disrupt communication, 'calmatives' to drug one's adversaries, sonic devices to confuse them, the spread of glues to inhibit their movement, and information warfare is itself part of the non-lethal arsenal. However, all three reasons for a reduction in violence that were outlined above are highly contentious. Information warfare can be used to undermine and suppress legitimate democratic debate as well as the subversive activity of terrorists, a fact admitted by both radical and mainstream commentators (Toffler and Toffler 1993, Arquilla and Ronfeldt 1997, Virilio 2002). Even the most ardent proponents of disruptive weapons, such as the Tofflers, concede that in reality the bloodshed of the battlefield is still horrific. The so-called 'surgical strikes' carried out by allied forces during the Balkans conflict inflicted a horrific 25% civilian casualties, according to NATO's own estimates. The first Gulf War has been described by the Tofflers as a 'dual war', on the one hand a clean information war, some of which was shown to us on CNN, and on the other hand an industrialized slaughter that was hidden from an audience that might otherwise have turned squeamish in the face of such horrors.

Writing about the Red Brigade in the 1970s, Umberto Eco observed that although modern terrorism claimed inspiration from Marx, its real influences could be found in science fiction and the cybernetic theory of Norbert Wiener (Eco 1977). The argument runs that one cannot destroy a factory by killing its owner, who would simply be replaced, one must instead feed the system aberrant information to confuse it and destroy it. According to Eco (1977: 179), 'The problem is that it [terrorism] has not drawn profound enough lessons on, nor has it studied cybernetics with sufficient care'. With the advent of information warfare, one might say this lesson has at last been taken to heart. Information warfare is less a part of a reduction in violence than an important feature of the changing face of violence. Information warfare is very much a part of what has come to be known as the 'War on Terror'. Current events are seeing a polarization between the so-called 'civilized world' as opposed to the terrorists and their supporters. This war has been defined by US President George Bush Jr as a never-ending war, which extends to all parts of the globe. Information warfare is a key doctrine in the development of the 'omnicrisis' that has been described in the speeches of G.W. Bush. The idea of a never-ending omnicrisis represents a blurring of the boundaries between the notions of attack and defence. The 'war on terror' was first used as a justification for the bombing of Afghanistan, which incidentally killed nearly double the number of civilians that were killed in the attack on the World Trade Center (*Observer* 14 July 2002), to say nothing of military personnel and conscripts killed. And in 2003 it was used as a pretext for an invasion of Iraq. If peace really is the objective of the 'war on terror' then it is difficult to explain why the chief exponents of this war are among the largest armaments suppliers in the world. Amnesty International has observed with regret that the five permanent members of the UN security council are also the world's largest producers and exporters of armaments. Paul Virilio has suggested a novel solution to the problem of terrorism which is both simple and well within the capacity of nations to carry out – outlaw the arms trade. He believes that if the weapons industry were stopped there would be no more terrorists (Virilio 1997a: 111). This may be a somewhat over-optimistic conclusion but, at the very least, the opportunities for terrorism would be severely restricted. This remarkable solution would also avoid the problem of 'blowback', which the CIA has identified as being a serious flaw in many of its previous operations (*Guardian* 15 September 2001). Blowback is precisely what happened in Afghanistan where extremism was actively encouraged by the CIA and ISI (the Pakistani Intelligence Service) when it suited their aims in the 1980s and 1990s. Early signs of future disobedience to their Western patrons could be detected when the Afghan warlords refused to return the high-tech Stinger missiles supplied to them by the US during their campaign against the Soviet Army. With the Soviet withdrawal, the country was abandoned to cope with ideological extremism, religious oppression, a flourishing drugs trade and bloody rule by competing warlords, all of which had direct funding from Saudi Arabia and Pakistan with the approval of Western governments (Rashid 2001). Over the past five decades, many nasty dictators have been the recipients of Western largesse when it suited these powers, only to be condemned later when they became an inconvenience. This unsavoury list includes the likes of Pol Pot, whose Khmer Rouge received support from the Thatcher government among others, Saddam Hussein who received armaments and support from Russia and the USA, and a score of Latin American dictators, from Pinochet to Noriega, all of whom had previously received significant backing from the US government. The fighters of the Taliban and bin Laden himself represent the latest additions to this list who, in their former guise as the Mujaheddin, had previously received the full endorsement of the US (Kolko 2002, Rashid 2001).[9]

Although not lethal in itself, information warfare does not herald the dawn of a new age of non-violence or a new 'peaceform' as the Tofflers have termed it. Information warfare represents a change in the nature of violence, it concerns the development of techniques for defending and attacking not only the corporeal body when playing a part in electronic warfare, but the very culture and identity of one's adversaries, such as is clearly the case with netwars.

## Maintaining identity and identity attacks

An important aspect of information warfare that tends to be underplayed in the existing analyses of the concept is the relationship with identity. Many commentators have skirted around this topic in their discussions of propaganda (Arquilla and Ronfeldt 1997, Toffler and Toffler 1993, Denning 1999). Arquilla and Ronfeldt put identity at the centre of their theory of information warfare when they define it as 'the use of information to impose one's will upon an adversary' (Arquilla and Ronfeldt 1997a: 14). However, they do not pursue the question of identity beyond this initial statement. Denning's theory of information warfare is also grounded in identity, as illustrated by her definition of netwars as an attempt to 'disrupt, damage, or modify what a target population knows or thinks it knows about itself and the world around it' (Denning 1999: 73). Information warfare concerns one's identity in two important respects, in the first place the construction of an identity may be crucial in prosecuting information warfare, as will be discussed below with respect to the Zapatistas. In the second place, one's orientation in the world can be directly threatened by information attacks such as identity theft, propaganda, public relations and psychological operations.

Some fascinating work on the relationship between information and identity can be found in the literature on cybernetics. Stafford Beer's 'viable systems model' of organizations states that identity is essential to the survival of an organization, and threats to its identity are threats to its very survival (Beer 1979). The identity of a system is an emergent property of the whole, it remains even if all the various parts of the system come to be replaced over time. The creation of an effective identity is necessary for a system 'to act as a unified whole' (Beer 1979: 418). It is precisely action of this kind that much information warfare is directed against. As has been explained previously, the techniques of information warfare are very much concerned with disruption, or as Virilio (2000a: 49) terms it, the 'remote guidance of confusion'. Beer was a visionary in pioneering the application of cybernetic principles to social organization. In fact, he worked briefly for President Allende in Chile in the 1970s to engineer the Chilean economy along cybernetic principles. Alas, this project failed with the CIA sponsored coup in which Allende was assassinated and the military dictator, Augusto Pinochet installed in his place.

The RAND Corporation's definition of information warfare, 'the use of information to impose one's will upon an adversary', provides an interesting comparison with the Zapatistas' own much subtler understanding. For a start, the Zapatistas have developed a communal identity, which not only involves the will of the local rebels, but has taken to the streets in demonstrations across the globe. They do not simply wish to take power and to impose their will, but to open up a democratic space in which different voices can be heard. In one of his many letters, Marcos compares the Zapatistas with marginal groups in music culture, like 'punks' and 'rappers'; these groups do not want to impose their identity upon everyone else, they merely seek for a space to be heard. This kind of information warfare is not so much about imposing one's will on another, as it is about resisting the impositions of the powerful and creating space for a plurality of wills to be expressed. In the words of the Sup,[10] 'We are not those who aspire to take power and then impose the way and the word. We will not be' (Marcos 2001: 159). This line is part of an extended definition of the Zapatista spirit, in which the phrase 'we will not be' is repeated many times over. In this simple phrase is shown a great deal of sensitivity towards the perils of assuming power and the danger of instating a tyranny of their own. As such they define themselves almost by a lack of identity, the very fact that the poor that make up their ranks have never been important enough to warrant a name in history.

The question of identity is very much at the heart of the Zapatista rebellion in Mexico, which mixes myth and history in an attempt to form an identity to mobilize resistance against the existing conditions of oppression and misery. These rebels have drawn upon Mexican history using the name of the rebel Emiliano Zapata to unite diverse indigenous peoples who have decided to join the struggle: 'With this new name, we name the nameless' (Marcos 2001: 20). The faceless peasants of the Chiapas mountains were frustrated when confronted with the power of the giant corporations and the devastating economic affects of North American Free Trade Agreement on their way of life. These peasants have struggled to assert their identity and communicate with the uncaring powers. The feeling of impotence and their frustration at being ignored is emphasized repeatedly in Subcomandante Marcos' writings. In one letter, Marcos compares the Zapatista predicament with that of Alice in *Through the Looking Glass*: '"Well, it's no use your talking about waking him," said Tweedledum, "when you're only one of the things in his dream. You know very well you're not real." "I am real!" said Alice, and began to cry' (Marcos 2001: 186). These indigenous people of Chiapas do not suffer merely from material deprivation, but from attacks on their culture, their identity and their ability to make themselves heard. The Zapatistas are made up from and represent the most impoverished members of the Mexican society, who are both 'nameless and faceless' as far as the governing powers are concerned. The rebels have exploited this anonymity in their use of the ski mask, which has a very complex role in the formation of the Zapatista identity. The black mask makes the Zapatistas faceless too, and as such provides a direct bond with the faceless peasants they represent, and who make up the majority of their ranks. The mere fact of attempting to hide their faces brings them to the attention of the authorities: 'We use black ski masks to show our faces. Only then can we be seen and heard' (Marcos 2001: 193). So, on the one hand, masks serve to hide their individual identities, but on the other hand they play an important feature of their powerful group identity. The use of masks has a long history in this conflict being deeply rooted in the local Mayan culture.

Identity is also inherently bound up with questions of authority and access. Electronic information systems often require the creation of a virtual identity to protect the integrity of data held in that system, and prevent access by the unauthorized. Denning (1999) places security at the centre of her theory of information warfare, because security is essential for the operation of financial and other information systems. Identity theft is a relatively common form of information warfare, especially credit card fraud. Playing with masks is very much the concern of the hacker who wishes to avoid identification. Guerrillas and disempowered groups tend to play with masks, creating their own ephemeral identities to mobilize support and move with a minimum of interference from the authorities. The state and powerful corporations work by fixing identities and thereby tracing and regulating movement. Identity is both a grammatical convention and a historical and cultural formation. Nietzsche's essay on the genealogy of morality suggests that, historically, identity was invented for the purposes of pronouncing guilt and exacting punishment (Nietzsche 1994). This is precisely the case of the netwar waged by the Zapatistas against the Mexican government. The Zapatistas wore ski masks to hide their identities, which the government were desperate to expose so they could subject them to the state's juridical procedures, and counter their efforts at resistance. This process of identification and unmasking was a key part of the state's information war against the rebels. Unmasking was crucial to destroying the power of mystique presented by their charismatic spokesman, Subcomandante Marcos. When the identity of Subcomandante Marcos was finally unmasked, he was discovered to be the son of a furniture salesman. The *New York Times* reported that, 'The moment that Marcos was identified and his photo shown and everyone saw who he was, much of his importance as a symbol vanished' (quoted in Taussig 1999: 236). Although the Zapatistas were extremely successful in forging their own identity and countering the strategies of information warfare used against them by the state, there is a sense in which the resistant identity loses its grounding in local struggles to communicate transversally across cultures and histories. As one commentator observed, 'Marcos is gay in San Francisco, black in South Africa, an Asian in Europe, a Chicano in San Ysidro, an anarchist in Spain, a Palestinian in Israel, a Mayan Indian in the streets of San Christobel. . .' (Taussig 1999: 264). The true object of information warfare is not only the information resources that a person or organization possesses but their very identity.

## A review of aims, methods and techniques

Information warfare is developing as a distinctive doctrine as a consequence of the revolution in information technology and the huge importance of information to the post-industrial economy. Broadly speaking, the targets of this kind of warfare fall into two categories, either information resources and infrastructure, or the ideas and belief structures that people use to orientate themselves in the world. This chapter has shown that civilians are often the agents of information warfare, especially in the kind of 'netwars' described by Arquilla and Ronfeldt (1997). The concept of netwar already highlights a blurring of the boundaries between the military and the civilian. Information warfare is only now coming to be introduced to the business world as a distinctive concept. One of the most well established forms of information warfare is that of propaganda and its civilian offshoot, public relations. But new techniques of information warfare are being developed, such as identity theft, viruses and email bombs. These mechanisms have a major impact on the way in which information is processed and disseminated throughout the globe.

This chapter has shown that the concept of information warfare has been explored in diverse areas of study, including military strategy, computing, business strategy, social theory and international relations. The argument has attempted to synthesize the major elements of information warfare, which revolve around intelligence, propaganda, public relations, speed, disruption and identity. This argument highlighted the importance of the material conditions within which information and ideas may thrive and spread, and the central role of communication technology in the development of information warfare. In the Table 4.1 the concept of information warfare is summarized in terms of three general elements: (i) the object of the warfare, (ii) the different methods for pursuing that object, and (iii) the specific technology that is used.

Network organizations have developed strategies that are very much concerned with their information resources and communications technologies, and have refined techniques for protecting these resources and, sometimes, carrying out offensive attacks against other organizations. In doing this they use a range of both high- and low-technology weapons, as outlined in the above table. These are related to the principles of nomadic strategy that were developed in Chapter Two. Nomadic strategy was defined in terms of the principles of speed, communication, leadership, friendship, the pack, the idea, peace, and calculation. These principles can be seen to underlie many of the techniques and methods discussed in the present chapter. For example, the principle of the idea is clearly embodied in the use of propaganda, public relations, slogans and rumours to achieve one's ends. In addition, the principle of communication is embodied in the techniques of disruption discussed in this chapter, where the ability to disrupt an adversary's communication system is becoming a primary

*Table 4.1* What is information warfare?

| | |
|---|---|
| Object of attack/defence | Information resources<br>Identity creation/maintenance |
| Methods | Disruption<br>Speed<br>Propaganda<br>Public relations<br>Intelligence |
| High-tech weapons | Internet and computer technology<br>Websites<br>Satellite networks<br>Communication prosthetics<br>Hacking and identity theft<br>Computer virus, email bombs, virtual sit-ins |
| Low-tech weapons | Pamphlets, leaflets, social engineering<br>Rumours, speech acts, slogans |

consideration in both offensive and defense operations. The importance of the principle of speed to information warfare is such that it has an entire section devoted to it here. Not all the principles of guerrilla strategy are clearly represented in this chapter, but it is nonetheless accurate to say that guerrilla strategy is particularly suited to this kind of warfare given the emphasis that such a strategy places on the power of ideas, networks of communication and the role of disruptive tactics.

What distinguishes information warfare as a concept is not any single technique, but the fact that we have crossed a technological threshold, where much of our social and economic infrastructure is based around information technology. One of the underlying themes that has recurred throughout this chapter is that techniques of information warfare tend to be applied differently when used by large powerful organizations, such as business corporations and states, compared with less well resourced bands of people, such as hackers, guerrilla fighters, or online communities. To some extent, the larger organizations are increasingly imitating the networks being developed by smaller nomadic bands. This can be seen in the military, where the influential think tank, the RAND Corporation, is advocating the use of networks by the state in order to counter the illegal networks set up by guerrillas or drug lords (Arquilla and Ronfeldt 1997). But this is also apparent in the civilian world and has been observed in the successes of the Linux community in transforming the software industry, and by groups such as Napster in challenging and transforming the music industry. These nomad networks use the speed of communication to disseminate information as widely and as quickly as possible and to evade regulation. In contrast, large corporations tend to slow the movement of information down, for example by using Intellectual Property Rights. This explains precisely the difference in tenor between the two preceding chapters. Chapter Two focused mainly on network organizations that had developed a nomadic form of strategy, and used information warfare and speed of communication to protect their communal identity. Chapter Three then focused on the use of networks by companies to halt the nomadic tendencies present in information, using mechanisms such as knowledge management and Intellectual Property Rights to capture and regulate the speed of the dissemination and usage of information. The main conclusion that should be drawn from this chapter is the need for individuals and organizations to become aware of the increasing threat of information warfare operations that may be pursued against them.

# 5 The virus metaphor

## Contamination and containment

> The age of globalization is the age of universal contamination.
>
> (Hardt and Negri 2000: 136)

> Worldly friendship is epidemic: everyone catches it, like a disease.
>
> (Barthes 1979: 138)

The local supermarket store sells oranges that were grown on a farm thousands of miles way, but the very same networks that have facilitated this global trade have also facilitated the spread of disease. Malarial mosquitoes have begun to spread outwards from their origins in the tropics, making a temporary home for themselves in the worldwide trade in rubber tyres, before finding a new home on foreign shores. Networks have enabled the spread of viruses, whether they are biological diseases or disruptive computer codes, and, to some extent, these same networks have been adapted to ward off the damaging effects of these viruses. The virus has already been discussed as a key weapon in the armoury of information warfare; however, the implications of the viral metaphor are more far-reaching. The virus has evolved as a principle of disorganization in step with the evolution of networks. The argument of this chapter also draws upon the virus metaphor to explain how changes in ideas and social relations are spreading throughout the network society. This viral transmission of ideas can be used to explain the phenomenon of cultural and linguistic homogenization that has appeared as a consequence of the globalization of communication networks and trade. The network must be understood as a dominant metaphor for describing the changes that organizations have been undergoing over the past few decades, where alongside the network viruses are also thriving.

### The power of the virus metaphor

A virus is a parasite, depending on a host organism for its source of energy and a site of reproduction. It has many unusual qualities; it hides its true nature from the host organism to gain entry, and once inside it makes further copies of itself (Smith 1963). Viruses often cause behaviours in the host organism that facilitate their spread, for instance, the common cold causes sneezing which projects viruses *en mass* through the air towards other nearby hosts. The rabies virus induces a kind of rage that leads the infected host to attack other animals and pass on the disease through the blood and saliva. Effective viruses are defined by the fact that they tend to be highly contagious. They are difficult to treat, not least because they can mutate rapidly, even travelling between different species. Viruses can kill the host cell or the entire organism, but killing the host on which it is entirely dependent is not an effective survival strategy in the long term. Many viruses can coexist without great harm to the host organism, even maintaining a permanent existence with the host.

Microbiologists have recently discovered that viruses can actually transfer genetic material between organisms of the same species and, in some cases, between different species (Margulis and Sagan 1997). On the microbial level we are swapping genetic information all the time. This fact has been exploited by medical science, which uses the virus as a major tool for the purposes of genetic engineering. Harmless viruses can be used to transfer healthy genetic material into a diseased body as part of a process of gene therapy (*Observer* 22 June 2003). The virus challenges traditional distinctions concerning the boundary of the organism. We form a rhizome with our viruses, which open our bodies onto the genetic material of life on Earth as a whole (Deleuze and Guattari 1988). The organization theorists, Carter and Jackson (2000) have proposed that organizations should be seen in this way, as rhizomes that have fluid boundaries, always subject to random growth and mutation. The current chapter will extend this analogy much further, with particular focus on the metaphor of the virus. Viruses will be shown to have important implications for the way in which we organize ourselves, and this is not simply metaphorical. The virus is a genuine force for disruption and disorganization. Viruses have no metabolism of their own and they are not classified as cells. They contain genetic material and they can be defined in terms of their main function, which is the expression of the information contained in their genome (Mims and White 1984). Seen as invasive parasites, disruptive of the host organism, they are purveyors of alien genetic information and are a perfect example of information warfare. To some extent, organizations are designed to ward off the effects of these alien forces, and it is to a brief history of this that we now turn.

## The network of disease: biological contamination

Fear of pandemics is now a media event through news coverage of diseases such as the flu, SARS and AIDS and in the sensationalized storylines of recent Hollywood disaster movies. Remarkably, much of the initial spread of AIDS in North America was traced to a single individual, Gaetan Dugas. Dugas was a flight attendant who travelled widely and had a surprisingly wide network of sexual partners throughout the US. In April 1982, of the 248 people then diagnosed with AIDS 40 could be traced directly to this one person. By his own estimates, he had around 250 different sexual partners a year and he may have infected thousands. Large ports, whether airports or seaports, serve as nodes in the global communication of disease as well as the global exchange of people and commodities. The spread of SARS around the globe has been traced to a single Chinese doctor who had become infected, whilst staying in Hong Kong. He subsequently passed the disease to 12 other guests of the Metropole Hotel where the doctor happened to be staying. Two of these guests remained in Hong Kong but the rest went on to five other countries across three continents. Within months, over 3000 cases of the disease had been reported across the globe. As a result, large cities, such as Beijing and Toronto, were placed under temporary quarantine. The appearance of pandemics is by no means a new phenomenon, but the rapidity with which the contagion spreads is accelerating in pace with globalization. In 1918, the spread of the Asian flu was greatly facilitated by the global movement of troops and equipment as a consequence of the First World War; it is estimated that this flu claimed the lives of between 20 to 40 million people around the world. Health is no longer merely the concern of national governments but has become a concern that is coordinated on an international level by institutions such as the World Health Organization and the Global Outbreak Alert and Response Network. In 2003, the UK government's health secretary, Alan Millburn, urged the WHO to take greater responsibility for global outbreaks of disease and to perform a regular audit of the ability of countries to deal with such epidemics. The instruments of medical control are increasingly taking an international dimension, where the world must be policed for dangers that are a threat to life itself.

Mechanisms of social control have been very much concerned with the issue of disease and contagion for centuries. When Michel Foucault described the history of disciplinary power, he traced its first major development to the outbreak of the plague (Foucault 1977). When the plague hit Europe in the seventeenth century, the inhabitants of cities were often subjected to the most severe regulations to safeguard public health. Streets and houses were placed under strict surveillance, and the comings and goings of inhabitants strictly circumscribed. Each household had to give a regular account of itself to a designated authority. The penalty for flouting such regulations included death. Quarantined households were also subjected to various forms of purification, by burning household contents and sprinkling perfume to purify the bad air. The registration of disease became increasingly centralized; the disorder and confusion of the plague was met with the order and discipline of a centralized administrative apparatus. Foucault saw the origin of disciplinary society in the plague, 'The plague ... is the trial in the course of which one may define ideally the exercise of disciplinary power' (Foucault 1977: 198). The techniques of confinement and surveillance that were developed for the administration of cities beset by plague were described by Foucault as the 'discipline-blockade' (Foucault 1977: 209). A century and a half after the plague, these mechanisms of social organization were to undergo another transformation with the creation of Jeremy Bentham's Panopticon.

The author William S. Burroughs has created a whole literature based around the concept of the virus. His work emphasizes how our lives are organized around diseases. New forms of disease are not merely the product of evolutionary chance, but are being manufactured within our laboratories as part of both medical and military research. He pointed to the *US Army Military Review* which contained discussions of the most appalling genetic research into 'ethnic weapons', that can be targeted against people with certain genetic traits (Burroughs 1986: 141). In 2000, a report entitled *Rebuilding America's Defenses* published by an influential neo-conservative think tank in the US, speculated on the use of such genetic weapons as 'a politically useful tool' (*The Guardian* 6 September 2003). The virus is a powerful force for disorganization and the extent to which this power can be domesticated is highly questionable.

Burroughs understood that we are on the brink of a biological revolution, the consequences of which can only be guessed, and which will usher in new demands for social control. Quoting from another newspaper article, this time in the *Herald Tribune* (June 1970) Burroughs (1986: 142) wrote: 'Dr Har Khorana at the University of Wisconsin has created an artificial gene. News that may rank with the splitting of the atom as a milestone in our control – or is it lack of control? – of the physical universe...'. He was deeply concerned about mankind's puerile efforts at tinkering with nature. Man appears to be obsessed with technological mastery, control and secrecy, yet history is replete with examples that highlight the futility of these efforts. Burroughs pointed to a report on the creation of a new cancer virus that had been created accidentally in the laboratory (*The London Times* 18 April 1971). These early concerns about the apparent lack of control in such genetic experimentation are now materializing in reality. Research published in the journal *Nature* (November 2001) has highlighted that experiments in genetic modification are by no means fully controlled or isolated to the laboratory. This article has shown that DNA from genetically modified maize had already cross-contaminated wild maize growing in the remote mountainous areas of Oaxaca, Mexico, in a matter of only a few years. The methods of transversal communication that are to be found in nature have transformed the genetic revolution from a laboratory study to a planetary wide experiment. The National Health Service in the UK has become concerned by the creation of 'superbugs', which are highly resistant diseases that have developed immunity to a range of treatments. These superbugs have developed within hospitals where, over time, diseases have become enclosed and concentrated. One of the main reasons for the development of resistant superbugs is that bacteria can actually benefit from viral infection. Viruses not only spread disease between people, but they also serve to spread immunity to drugs between bacteria (Margulis and Sagan 1986). By the mechanism of 'transduction' – that is, the sharing of DNA via viral infection – bacteria can rapidly adapt to new antibiotics and spread their immunity to other strains of bacteria. Transduction allows bacteria quickly to modulate their form to suit even the most hostile of environments. This means that immunity is developed, not just from random mutation, but because bacteria share their immunity between each other. According to the microbiologist Lynn Margulis, it has been estimated that such immunity from penicillin would have taken 'about a million years for random mutations to produce it' (Margulis and Sagan 1986: 91). Transversal communication in nature, particularly at the microbial level, suggests that the virus will always be a powerful force for mutation and what might be seen as a force for disorganization, at least from the perspective of the administrator.

## The network of machines: the contamination of computers

The dawn of the information age and the computer revolution was accompanied by the onset of a new plague – a disease that can infect machines themselves. A computer virus is a piece of computer code that attaches itself to an existing program, where it is activated at the same time as the host program is activated. When this happens the virus may begin to make copies of its own code in previously uninfected files. It may also execute other instructions, such as to delete files on the hard drive or to spread itself to other machines, which it does by copying itself to a floppy disk when one is present or by attaching copies of itself to outgoing emails. Viruses need not be programmed to delete files to cause damage to a network. Indeed, much damage can be done merely by the massive increase in email traffic they can generate. The 'Love Bug' virus, which circulated the Web in May 2000, is estimated to have caused £5 billion pounds of damage and disruption to business (*Guardian* 24 August 2003). The damage done by computer viruses is not purely economic and, on occasion, they can become a serious threat to human safety – if they infect the systems of organizations such as nuclear power plants, hospitals or national railways. For instance, in 2003, the Davis-Besse nuclear power plant in Ohio had to be shut down temporarily after its safety monitoring systems had been compromised by a computer virus.

The term 'virus' was first used in relation to computer machinery by Gregory Benford in 1970 to refer to unwanted computer code that could self-reproduce on what was then the ARPANET. By 1998, more than 130,000 kinds of computer virus had been created (Denning 1999). In that same year, the FBI Computer Crime and Security Essay reported that nearly three quarters of the organizations they surveyed had detected viruses within the past year. Estimated damage from such infections ranged from between $50 and $2 million. At present, however, most of the viruses in circulation have no other aim than their own propagation and do no further damage. A virus usually requires the cooperation of the user of the host computer system to activate itself. Typically the user must open an email attachment where the virus resides to activate the virus code. Viruses gain entry to host systems by means of stealth, they remain hidden from detection often by using some form of encryption. They do this to avoid being deleted, and also to allow for a period of time during which they can spread before they carry out any further instructions. One of the most cunning forms of virus is hardly a virus at all, and functions merely as an email. The email itself appears as a warning of a virus already present on the user's system and demands the user delete the offending piece of code immediately. In actuality, the specified code is a genuine part of the computer's operating system, and by deleting it, the user ends up causing damage to their own computer system. This example shows that there is a close tie between a computer virus and language itself, both having the capacity to effectuate commands on a host computer system. The 'I Love You' virus was probably the most effective and damaging virus to infect the Internet since its inception. It has been estimated that it destroyed data on about 45 million computers around the world. This virus was particularly successful not only because of its code, but also because of the mechanism by which it spread, on an email entitled 'I love you'. This kind of virus, known as a 'worm', has many variants, and depends largely upon the power of its name for its effect. Common names include, 'I LOVE YOU', 'Joke', 'Mothers Day Order Confirmation', 'Dangerous Virus Warning', 'Virus Alert', 'Important! Read carefully!' Dressed as good ideas, these viruses know how to market themselves. They literally command you to open them, and thereby become infected. The idea of love itself seems to be a highly contagious virus. The French semiologist Roland Barthes (1979) made the same observation some years before the Internet took off in his book on a lover's discourse. In this work, Barthes described the process by which groups of friends gossip about love and pass information between each other as being 'like a disease' (Barthes 1979: 139).

## Pathological identities and moral diseases

A central motif throughout Michel Foucault's work is the medicalization of social issues, and how criminality and deviance in general came to be seen as diseases. Criminal, sexual and mental delinquents are confined, not simply to treat them, but to protect the outside population from possible contamination. Foucault's work focused on the history of confinement in terms of how it facilitates supervision and treatment, but also in terms of how it prevents communication and the further spread of disease. A 'great fear' emerged as places of confinement were seen not simply as containers of madness, sickness and crime, but as the breeding ground for these diseases (Foucault 1971). In the second half of the eighteenth century, this fear led to the greater segregation of these institutions from the outside world. As Foucault (1971: 206–207) himself observed, 'The hospital, the house of correction, all the places of confinement, were to be more completely isolated, surrounded by a purer air: this period produced a whole literature concerning the airing of hospitals, which tentatively approaches the medical problem of contagion, but aims more specifically at themes of moral communication'.

In his essay, *About the Concept of the Dangerous Individual*, Foucault highlighted the historical development of how a criminal identity became conceived separately from the crime itself (Foucault 2000a). Foucault's essay examines perceived deficiencies in the legal apparatus and the mechanisms that were created to cope with these deficiencies. One of the most serious issues he identified in the history of the legal system was what happened when an accused refused to defend their actions in court, especially in the case where monstrous crimes were being tried. These confessions were considered an essential part of legal rationality, because it was here that the accused explained and accepted their guilt. What could the court do where no such explanation was forthcoming? Gradually, many of these cases, particularly the most extreme examples, were explained in terms of madness. The most irrational aspects of criminal behaviour were explained away as a disease of the mind, a deviation in their being. During the nineteenth century, there was a gradual pathologization of crime and the growth of an expert profession of criminal psychiatrists, whereby crime became defined as a disease that could be treated. On the one hand, crime was increasingly seen as a disease of the mind that could be treated for illness, and on the other hand, it was seen as a disease of the social body that needed to be protected from such delinquency. Punishment was not simply exacted in accordance with the severity of the crime, but to protect the social body itself from contamination from the 'dangerous individual'. The criminal was defined both as someone who had been found guilty of committing a crime and whose very nature was itself criminal.

The treatment of specific kinds of identity in pathological terms also extended to the ideas of madness and of sexuality. Foucault's work on sexuality traces the historical processes whereby sex became seen less in terms of an act and more in terms of an identity. Whereas the concept of sex can be defined within the field of biology, the concept of sexuality has been defined within the field of the human sciences. Deviance, such as homosexuality, was understood as being both immoral and indicative of a psychological pathology. During the eighteenth and nineteenth centuries there was a growth in the classification of these diseases, the treatment of which was associated with the growth of the disciplinary apparatus in Europe (Foucault 1977, 1981). The twentieth century has seen its own revolution in the mechanisms of social control, and many of these are explored in the writings of William Burroughs in both his fiction and non-fiction essays. Like Foucault, Burroughs was very much concerned with the relationship between identity, the body, and the exercise of power. Burroughs was a homosexual and a drug user, both of which were marginalized identities subject to repression by the police and the legal system. Burroughs noted that it was not his activities as such that were persecuted by the system, but his very identity. He highlighted the fact that, in certain states in the US, such as Louisiana and California, being an addict was a crime in itself (Burroughs and Odier 1989: 147). It was not just manifest behaviour that had become outlawed but specific 'deviant' identities.

Both Foucault and Burroughs described the various ways in which organizations normalize identity and treat deviant identities as if they were diseases. The concepts of criminality, madness and deviant sexuality have all come to be associated with a person's identity. They are all judged according to some criteria of health, an idea of normality to which they must conform. If criminality, madness and deviant sexuality can be understood as diseased identities, then Burroughs argued that identity itself could be seen as a disease, an infection of the human organism. According to Burroughs, the concept of identity is nothing more than a 'word virus'. In his work he developed a brilliant and highly persuasive parody of the theme of contamination, where individualism and morality are themselves the symptoms of disease. Burroughs used the metaphor of the virus to develop a highly sophisticated critique of society and its methods of control. He described language as a virus that has spread by infecting the human animal. It is worth quoting his summary of this unusual theory of language at length:

> My general theory since 1971 has been that the Word is literally a virus, and that it has not been recognized as such because it has achieved a state of relatively stable symbiosis with its human host; that is to say, the Word Virus (the Other Half) has established itself so firmly as an accepted part of the human organism that it can now sneer at gangster viruses like smallpox and turn them in to the Pasteur institute. But the Word clearly bears the single identifying feature of the virus: it is an organism with no internal function other then to replicate itself.
>
> (Burroughs 1986: 47)

The virus gains entrance by fraud and maintains itself by force. An unwanted guest who makes you sick to look at is never good or

> beautiful. It is moreover a guest who always repeats itself word for word take for take.
>
> (Burroughs and Odier 1989: 189–190)

Burroughs analysed human history in terms of the relationship between the word virus and its human host. In fact, he stated that history is not really a history of humanity, as commonly understood, but a history of the word virus. He observed that modern man has lost the option of silence. It is simply impossible to stop sub-vocal speech, the continuous drone of your internal monologue. A person may struggle to attain such silence but when they do so they are always met with a great deal of internal resistance. This resistance is an organism that forces you to talk; it is the word virus (Burroughs 1968: 39). This virus can be found at the very start of Christian theology, 'In the beginning was the word...' (St. John 1:1). Jesus was the living embodiment of this virus – the word made flesh. Eastern religions such as Buddhism have been preoccupied with our relationship with words, specifically with respect to silencing one's internal monologue. The extent to which Burroughs sets himself against the word is difficult to determine, especially given his role as a writer and master wordsmith. According to Burroughs, the word is the most powerful form of social control ever invented, and much of his work is an apprenticeship in how to sabotage and disrupt this system of control. All kinds of techniques can and are used to influence our perceptions, our purchasing patterns and political choices. Colours, smells, music and words are all commonly exploited to influence our perceptions of our environment to sell goods in supermarkets and advertise them on TV. The shear range and ubiquity of these techniques leads Burroughs to say that Orwell's *Nineteen Eighty-Four* looks like a benevolent utopia in comparison (Burroughs 1986). However, of all the various methods of control, words remain the most potent, '...words are still the principle instrument of control. Suggestions are words. Persuasions are words. Orders are words. No control machine so far devised can operate without words, and any control machine which attempts to do so relying entirely on external force or entirely on physical control of the mind will soon encounter the limits of control' (Burroughs 1986: 117). The control mechanisms of the word virus have been intensified in the twentieth century with the emergence of the mass media. The remaining sections of this chapter will further explore the viral properties of language and communication, and their relationship to organization and control.

## Networks of words: the contamination of language

One of the most peculiar ideas to appear on the Internet in the year 2000 was the notion of the 'ideavirus', developed by marketing executive Seth Godin. Godin argued that crazes for products mirror the behaviour of diseases, especially venereal disease, in the ways that they are spread by a promiscuous and energetic few. Once the ideavirus has taken hold in a small population it can then grow exponentially. To demonstrate his point, Godin released his book, *Unleashing the Ideavirus*, free on the Web, and within a few months of appearing in August 2000 it had been downloaded by over a million people (*Observer* 26 November 2000). Godin believed that traditional advertising would disappear as the new media takes over, and that word-of-mouth recommendations through the new media will replace advertising. Examples of this kind of marketing can already be seen in companies such as Amazon.com, which ranks the products it sells using its own customers, and even ranks the reviewers as well. The ideavirus is really just a technological variation of traditional word-of-mouth marketing, but this is greatly amplified by the power of new communications networks. In this new kind of marketing it will be important to identify the 'sneezers' who spread the word virus, and fertile places for the spread of contagion, such as schools or student unions (mirroring real viruses). Godin's approach might be seen as marketing hype, but he is by no means the only person to see an analogy between the spread of ideas and viruses. For example, Professors Hardt and Negri (2000) have explained the movement of revolutionary ideas, such as the slave revolts of the Caribbean, in terms of the virus. The revolutionary ideas modulated their form to survive and flourish on each different island and its particular material context.

Technology and trade have provided a powerful mechanism for the global spread of ideas. At the dawn of the electronic revolution, Marshall McLuhan made a claim for the emancipatory potential of the new technologies, stating his beliefs in somewhat religious tones, 'The computer, in short, promises by technology a Pentecostal condition of universal understanding and unity' (McLuhan 1964: 90). McLuhan envisaged a new era of collective harmony and peace arising naturally from these technological developments. Specifically he believed that computers would allow instant translation between different languages, and that this would enrich mutual understanding. What we know now with the emergence of the Internet is that things are not translated from any language into any other, but from any language into English. In fact, they are often not even translated by computer at all since English has become the language of globalization, commerce and the Internet. The information society has not only seen a rise in the quantity of information that is easily accessible, but has also seen a massive rationalization and standardization of this information. On the one hand, the Internet has seen a tendency towards the decentralization of information dispersed throughout its architecture. But, on the other hand, there has been an associated tendency for increased standardization and recentralization (Reinerman 2003: 249). New regimes of control are emerging for the regulation of flows of information, controlling the access rights to these flows. This recentralization is not simply a demand for concentrated power but is crucial in establishing the common standards that allow for the exchange of information in the first place. Centralization and standardization are as much the result of a pragmatic imperative of communication, as they are a desire for concentrated power.

In 1999, it was estimated that English was spoken for the first time by more people as a second language, than by native English speakers. A report on globalization by the telecommunications giant, British Telecom, is very revealing in this respect. This report found that languages are fast disappearing, being replaced instead by the language of commerce. BT is concerned about the effects of globalization as suggested by the following research it commissioned and then published on its own website:

> . . . up to half of the (approximately) 6,500 languages now spoken are already endangered or on the brink of extinction, and linguists estimate that a language dies somewhere in the world every two weeks. Much of the remaining linguistic diversity is carried by small communities of indigenous and minority people. In fact, some predict that we may lose over 90 per cent of the world's languages during the next century. Already, more than a quarter of the world's population – 1.7 billion people – now speak English.
>
> (www.bt.com/world/sus_dev/)

This section of BT's report concludes by stating somewhat blandly that, 'The negative aspects of cultural imperialism and intolerance could be replaced with an understanding of our world-wide impacts and intrinsic inter-linkages, which are underpinned by our own cultural uniqueness'. Although concerned, BT believes that the benefits of cultural imperialism outweigh the costs, with little need to do anything further about the matter. One might be tempted to suggest that this perspective is a little coloured by the nature of its business. The English word virus triumphs, contaminating more hosts than any other variant. It's good to talk – in English.

Historically, there has always been a close relationship between disease and language, the ancient cultures of what is now Latin America were destroyed as much by the alien language which infected their tongue as by the alien viruses which consumed their bodies. De Landa (1997) has noted that 50 years after its first encounter with Cortes, the population of Mexico had been reduced to a mere tenth of its former size. The conquistadors brought with them a different culture and a new language, but in their wake came smallpox, measles, flu and other viruses. Tens of millions of Amerindians were killed by diseases newly introduced to the continent and for which the native immune systems had no effective response. The biological plague and cultural plague travelled side by side. It is no coincidence that the first attempt at a book on modern grammar was published in Spain in 1492, the same year in which Columbus first set foot in the Americas. When the author of this work, Elio Antonio de Nebrija, explained its importance to Queen Isabella of Spain, he did so by saying that a standard language would help to facilitate the imposition of colonial culture and further enslave the barbarian tribes. In the words of Nebrija himself, 'Soon your Majesty will have placed her yoke upon many barbarians who speak outlandish tongues. By this, your victory, these people shall stand in a new need; the need for the laws the victor owes to the vanquished, and the need for the language we shall bring with us' (quoted in Illich and Sanders 1989: 68–69). A standardized tongue would be useful, so Nebrija argued, to the colonial apparatus. Furthermore, vernacular dialects and the language of the so-called barbarians were outside of state control and were thus a potential source of resistance and rebellion.

The power of vernacular language as a potential site of resistance to authority has also been noted by the eminent historian, Christopher Hill, in his study of the role of the bible during the English Revolution (Hill 1991). Latin was then the language of power and the means of expression of religious truth throughout much of Europe. It was in Latin that the priests and the elite classes were educated, and it was in Latin that expert knowledge was written down and circulated. Throughout the Middle Ages the bible was kept in Latin, and the truths contained therein could only be revealed to the common people through the intervention of an elite class of priests. This was not simply a matter of circumstance, for translation into the vernacular was absolutely forbidden by law. Hill also noted that the first English translations of the bible appeared during periods of great political turmoil, such as Wyclif's translation just prior to the Peasant Revolt of 1381. In the hands of the laity, the bible became a revolutionary text, and its translation into the vernacular spawned numerous sects, such as the Levellers, the Diggers, the Ranters, the Quakers, and the Shakers. During the civil war years in England, the 1640s, censorship collapsed and, as a result, there was an explosion of printed material, including several versions of the bible printed in English. Whereas no newspapers had been produced in the years prior to 1640, there were 700 in circulation only a few years later in 1645, and opinion formation was no longer the domain of an educated elite. This explosion of alternative ideas and printed material ended shortly thereafter with the restoration of Charles II to the throne. The translation of the bible was by no means the cause of the English Revolution, but it played a central role in the expression of heretical ideas and their control through censorship and the regulation of translations into the vernacular.

The control of language and the freedom of expression are implicitly related. This relationship is clearly recognized in the First Amendment of the US Constitution, which protects the freedom of the press. We have just read of a few historical examples that clearly illustrated attempts to regulate the expression of ideas at a national level. These techniques of control found their ideal literary expression in George Orwell's concept of

'Newspeak', which he outlined in the novel, *Nineteen Eighty-Four*. The engineering of language in Newspeak required a powerful central state, and so may not be particularly relevant to modern market orientated democracies. Nevertheless, a form of linguistic standardization and homogenization is taking place on a global scale, but through the mechanism of the market rather than through overt state interference. As the BT report revealed, the English language is spreading like a virus throughout these global networks of communication. To a large extent, this standardization can be understood as a pragmatic imperative of communication, making the exchange of information more convenient. However, it might be argued that linguistic standardization also represents a form of cultural imperialism. Linguistic diversity is not the only thing at stake with such attempts to regulate language, as will be demonstrated in the following analysis of business jargon and war propaganda.

## Contamination of language in business speak

The field of management, at least in its most populist form, provides some excellent examples of the abuse of language. In 2000, a popular game appeared on the Internet called 'Buzzword Bingo' or, alternatively, 'Bullshit Bingo'. The website gave a list of business jargon in the form of a bingo card, which could be printed out and used to enliven otherwise dull meetings. Once a line of buzzwords was completed, the winner was supposed to stand up and announce to his or her colleagues: 'Bullshit!' The list of terms is a vast and ever growing one, including: empowerment, playing hardball, synergy, results-driven, valued-added, benchmark, total quality, knowledge base, client focused, leverage, market driven, thinking outside the box, win–win, best practice, and on and on. The experts from the management consulting industry have emerged as the masters of this new art form. The cartoonist Scott Adams has founded an international reputation based on lampooning this kind of behaviour in his *Dilbert* cartoon strips. The Plain English Campaign itself has criticized the business world for being a prime channel for creating and spreading jargon and gobbledegook. At least one company, Colonial Financial Services, has taken this issue seriously and actively supports the Plain English Campaign in its efforts to help its customers decipher such obscure business speak (*Independent* 14 July 2000). Even mainstream analysts have taken the opportunity to criticize certain aspects of business language, which they believe can often hide the truth of what is really going on. Harvard economist Michael Porter (2001) has condemned the kind of jargon that emerged alongside the apparent success of the dot.com companies towards the end of the 1990s. He directed his attack against terms such as 'e-business', 'e-strategy', and 'business model' all of which were supposed to be a part of the 'new economy'. There was a proliferation of new jargon and accompanying images that led to the inflation of dot.com companies, which were in reality making very little revenue (de Cock *et al.* 2001). Other than the use of the prefix 'e' there was also much use of the adjective 'new', which was supposed to refer to new virtual markets and 'business models'. Slogans such as 'We're the dot in dot.com' and 'We put the "e" in strategy' were used to market these ideas. Many traditional firms also hopped onto the bandwagon, proposing new 'business solutions' as offered by IBM, or 'innovative solutions' as offered by Price Waterhouse Coopers. A host of images were also employed to illustrate these new terms, which emphasized youth, informality, and the speed and global reach of technology. However, the actual meaning of the 'e' and the importance of the '.' was never adequately defined or explained in this lexicon of jargon. After an extended period of excessive speculation and share price inflation, to say nothing of the outpouring of jargon, the market for these companies collapsed.

Management literature may, in general, be said to suffer from similar kinds of problems surrounding its ingenious use of language. One management writer, professor Gibson Burrell, has characterized this literature as 'Heathrow Organization Theory'. These books are generally a form of self-help manual crammed full of advice on how to get ahead in business or how to cope with the trials and tribulations of stressful or unfulfilling work, and they sell by the cartload in the airport lounges throughout the world. A relatively harmless diversion one might think, but perhaps symptomatic of a deeper problem. In fact, John Kay, Professor of Economics at the London School of Economics, hit out against these kinds of books in his column in the *Financial Times* in 2002. In particular, he looked at the recent management best-seller, *Who Moved My Cheese?* This book had sold over 10 million copies by the time it had come to Kay's attention. Despite its popularity, he attacked the book for the way it patronized its audience. The book advises its readers that they should focus on getting more cheese and not worry about the bigger issues, such as who moved their cheese. Those bigger issues are to be left to those better placed to know, who are higher up the corporate hierarchy. More worrying still is its use of a ridiculous analogy (cheese moving) and very poor stylistic quality. Using a piece of dedicated computer software, Kay estimated that the book assumes a reading age of a child with six years education. What concerned him most is that the book was quite typical of the genre. In Kay's words,

> It seems that no business publisher can underestimate his readers. What is depressing about *Who Moved My Cheese?* is not just that its promoters think that a business audience should be addressed in terms appropriate to a 10-year-old. It is that the success of the book appears to confirm that they are right. Other business books have similarly low aspirations.
>
> (*Financial Times* 7 May 2002)

In a more light-hearted vein, Kay stated that there is one notable exception to this trend – the consultancy industry. This industry is distinctive in its use of language that is far from easy to comprehend, being riddled with jargon and convoluted prose. This jargon marks out the terrain of the highly paid expert. Professor Kay would no doubt be heartened to discover that, since his article was published, the consulting industry has taken this issue to heart, in its own inimitable way. Early in 2003 the consulting firm, Deloitte, announced the release of a piece of software called the 'Bullfighter', which was designed specifically to prevent the recurrence of corporate disasters such as ENRON. In essence, this software acts as a kind of advanced spellchecker that examines documents, such as company reports, for the occurrence of what Deloitte describe as 'bull'. It has a picture of a bullfighter on it just to hit the point home. This war on bull began when Deloitte discovered that ENRON's company reports became saturated with jargon as it descended into chaos and financial ruin. This proposed software is an indictment of business language and the critical capacity of the business community, assuming as it does that a simple spellchecker will pick out 'bull' that a highly trained employee would not. Jargon is only a symptom of a deeper concern about the effects of language on thinking.

This thesis may be seen as an updated argument of what George Orwell (1948) termed 'Newspeak', a fictional form of the English language in which certain words were changed or removed to facilitate thought control. In his book *Nineteen Eighty-Four*, the citizens of the totalitarian state of Ingsoc were given an impoverished vocabulary with which to speak and were bombarded instead with slogans such as, 'War is peace. Freedom is slavery. Ignorance is strength'. But whereas Orwell's Newspeak was imposed by a centralized totalitarian government, the spread of English and business-speak is a function of both the success of information technology and the marketing methods of the huge capitalist enterprises which sell and use these technologies. Whereas the fictional Newspeak reduced the richness of language for expressly political purposes, the disappearance of cultural diversity is now rationalized in economic terms, where freedom is conceived largely in terms of the 'free market'.

## Contamination of language in war speak

During the Second World War, the novelist and political essayist George Orwell worked in the production of propaganda for the British government. He saw this as a necessary part of the fight against the fascist powers in Europe, but it also gave him much cause for concern, which subsequently he wrote about in political essays and fictional works such as *Animal Farm* and *Nineteen Eighty-Four*. He was wary of propagandist language and held the speeches of politicians in particular disdain, which he thought were, 'a mass of lies, evasions, folly, hatred and schizophrenia' (Orwell 1994: 357). Orwell was not simply worried that political discourse contained lies, but that it had insidious effects on language itself. If thought corrupts language by the use of hollow phrases to cover up lies, then language in turn can corrupt thought. He was so concerned about this problem, that he wrote an entire book on it, *Nineteen Eighty-Four*.

It has been said that the first casualty of war is the truth. Language itself is another casualty. *Nineteen Eighty-Four* describes a society in which language has been engineered and pared down to such an extent that heretical thought is not possible beyond the mere recognition that it is heretical. This can be seen to be the essence of much political rhetoric, for example President George W. Bush's claim that, 'You're either with us or with the terrorists'. Any possible criticism is warded off in advance with an implicit threat. The propaganda machine and its attack on language has been no more apparent than in the rhetoric concerning the Gulf War of 2003, where oil has brought about a clear convergence between strategic interests and business interests. During the coverage of this conflict the absence of any concrete information about the progress of the war was striking. This is not merely a matter of propaganda but involved the creation and use of a whole jargon of warfare. We saw images of soldiers firing guns in a desert somewhere, and lines of enemy soldiers handing themselves in to the victorious forces of the US by the barrack load. Instead of information, a number of slogans were repeated *ad nauseam* such as 'surgical strike', 'stiff resistance', 'air campaign', 'soften up defences', 'collateral damage', 'coalition forces', 'friendly fire' (now renamed 'blue on blue'), 'Weapons of Mass Destruction', 'liberation' and 'War on Terror'. These phrases carried little information in themselves and yet were repeated countless times on British television as the main part of the war coverage (or the 'War on Saddam' as the British ITN news termed it). Some phrases are easy to decipher, we know that collateral damage means dead and injured civilians, and that 'soften up defences' means the saturation bombing of enemy soldiers. Some terms are a little more resistant to interpretation, but with the benefit of hindsight we might be tempted to think that the 'coalition forces' were just the US and the UK, where uncooperative countries were either bribed to cooperate or subjected to espionage. Perhaps the biggest single attack on language was the phrase 'War on Terror' which makes no literal sense whatsoever. Terror is an emotion, intangible and impossible to isolate; perhaps this explains its rhetorical force. However, it remains unclear how a war can be waged against an emotion, not least given that the outcome of war is this very same emotion. It may simply be a mistake that the slogan was not the more accurate 'war on terrorism', but this conveys a very different image. It is just as impossible to fight a traditional war against terrorists who may well be indistinguishable from civilians, where there is a danger that war degenerates into a war against civilians. The idea of an all-pervasive terror, for which we must be constantly on our guard, provides a backdrop against which one can declare a state of emergency, even in the absence of a specific enemy. The ideals of 'freedom and democracy' have also suffered in the mouths of propagandists. These ideals are embodied in the legal constitutions of modern democracies explaining the rights of their citizens, but they are being increasingly suspended in the so-called 'war on terror'. There seems to be some irony in the fact that democratic rights are being suspended in many supposedly democratic countries so that they might be given to those who live without democracy. With unashamed honesty, *Fortune* magazine described the triumph of the war in terms of, 'Making Iraq Safe for Capitalism' (*Fortune* July 2003).

Arguably, the effect of propaganda during the Gulf War in 2003 was very limited, as witnessed by the massive anti-war protests held in London and other large cities throughout the world by those who were not persuaded by their leaders. However, even these protests declined when the war began in earnest, and there was a significant drop in opposition as reports came in of coalition successes. During the Gulf War II, there was a non-stop flow of stories about the discovery of possible weapons of mass destruction. Not a single case turned out to be genuine. Despite this, the propaganda was relentless. The lines of association had to be laid, 'WMD', 'terrorism', 'liberation' and so on. The lack of any such evidence continued to be a thorn in the UK Labour Government's side. Regardless of the rights and wrongs of the war, the extensive use of the slogans and propaganda seriously distorted understanding of events. This so-called news has limited information content, and its primary purpose is the equivalent of a general order commanding obedience. Back in 1946, Orwell expressed his concerns about this kind of political language in the following words:

> When one watches some tired hack on the platform mechanically repeating the familiar phrases – *bestial atrocities, iron heel, blood-stained tyranny, free peoples of the world, stand shoulder to shoulder* – one often has a curious feeling that one is not watching a live human being but some kind of dummy: a feeling which suddenly becomes stronger at moments when the light catches the speaker's spectacles and turns them into blank discs which seem to have no eyes behind them. And this is not altogether fanciful. A speaker who uses that kind of phraseology has gone some distance towards turning himself into a machine.
>
> (Orwell 1994: 356)

Orwell described how communication could be reduced to the transfer of information between the elements of a system, rather than a genuine dialogue that elicits critical reflection and discussion. The human mind becomes something that is dictated to, a mechanism that receives and obeys orders. As he explained in the appendix to *Nineteen Eighty-Four*,

'The intention was to make speech, and especially speech on any subject not ideologically neutral, as nearly as possible independent of consciousness' (Orwell 1948: 265). Orwell was not the only author concerned with such ideas. The Bullfighter software developed by Deloitte has the explicit aim of detecting jargon, but in a way that is totally divorced from a reflective human consciousness. The effort of detecting 'bull' is left to a machine, suggesting that consultants themselves were not up to the job. A few decades before Orwell, Edward Bernays was expounding the same kind of thing in terms of public relations. Bernays was a hugely influential figure in US industry and the news media in the 1920s and 1930s and is considered to be the founding father of public relations. Bernays (1923) went as far as to define the function of public relations as a means of reducing human communication to a simple matter of stimulus and response. The PR man 'must discover what the stimuli are to which public opinion responds most readily' (Bernays 1923: 96). The nephew of Sigmund Freud, he believed that human instincts were essentially fixed, but that people's opinions could be manipulated by means of the mass media and in this way human nature itself was 'subject to modification' (Bernays 1923: 150). The PR man does this by drawing on existing stereotypes and creating new stereotypes where necessary. These can be used to trigger the required response from the audience. One of the main ways in which control operates in the mass media is by 'laying down lines of association' (Burroughs and Odier 1989: 176). Once these lines have been laid, the media need not provide any evidence for its pronouncements, or justify itself to the consumer. It simply needs to provide the appropriate trigger. This is precisely how advertisements work and it is behind the whole concept of brand loyalty. Politicians and advertising executives have always relied heavily on this method. Just think of the number of times the names of Hitler and Churchill have been appealed to by politicians to justify a contentious policy, regardless of the intricacies involved. Indeed, these names were drawn on yet again in an attempt to polarize opinion during the Gulf War of 2003. Virtually every single advert on television draws on well researched stereotypes for its persuasive power; a man dressed in a white laboratory suit explains the health benefits of a particular brand of margarine or shampoo, an introverted geek attracts a beautiful women on applying a certain brand of deodorant. Bernays proposed the use of a range of stereotypes, including the manipulation of images, clichés, emblems and trademarks (Bernays 1923: 164). The kind of abuses of language described here should not be understood merely as attempts to change public opinion. They are attempts to reduce communication to a mechanical process of stimulus and response, where the mind is seen as a recording device for receiving orders rather than an active consciousness.

## Towards a programme of defensive measures

> If we recreate speech we will be able to resist.
>
> (Virilio 1999: 87)

Words are part of the oldest and most effective control system in human history; however, language is also the basis of human cooperation and community. Socialization occurs through language, and we should be reminded that, 'The best way to love one another is through language' (Virilio 1999: 65). If language can be attacked through the propaganda of the mass media and its communications networks, it can also be defended against these attacks. The necrosis of language that is apparent in business-speak and war-speak has had its opponents in both the mainstream business literature and more radical literature. Defensive measures can be taken against this mechanistic usage of language and the reduction of language to a set of orders or instructions. These countermeasures will be explored in the following two sections, the first of which looks into techniques for the disruption of the media monopoly and the proliferation of alternative channels. The second of which looks more closely into the metaphysical and viral presuppositions of language.

### *Disruption and the multiplication of channels*

Control works best when the intentions of the controllers are well hidden and imperceptible. Control often operates through monopolies and there have been many criticisms of the monopoly power of the media, most notably in the work of Noam Chomsky. In fact, the machinery of control can sometimes be turned against those in the seats of power. Control in ancient societies often depended on illiteracy and the specialist knowledge of the priest, but modern control depends upon literacy for its power: 'Modern control systems are predicated on universal literacy since they operate through the mass media – a very two-edged instrument, as Watergate has shown' (Burroughs 1986: 118). Literacy is the cornerstone of many modern forms of control, where communication has become the categorical imperative of the control society. Some years ago, this imperative was the subject of a British Telecom advertisement, which informed us that 'It's good to talk'. In fact, literacy is a two-edged sword, and Burroughs himself outlined a number of ways to undermine the centres of monopoly control. One such method is the 'cut-up', a technique that scrambles the original message of a text or broadcast. The cut-up works by dividing a message or text into sections and then rearranging them in a random order. The resulting cut-up introduces a random element, outside the cybernetic system of control; it distorts and disrupts media control by cutting the associational lines laid down by this media (Burroughs and Odier 1989: 176). Much of Burroughs' literary technique is premised on this simple idea, where he uses cut-ups for literary exploration and hones the results of these experiments for artistic style. The cut-up can be performed on all forms of communications where recording devices may be used, including written texts, taped sounds, conversations and video. The proliferation of different channels of communication and recording devices can serve to undermine the monopoly control of the mass media. Perhaps this idea has been most clearly demonstrated in the notorious case of Rodney King, whose vicious beating by police officers of the LAPD was captured on home video, the broadcast of which provoked the massive riots in Los Angeles. The media monopoly still exists, but technological changes that offer a supplementary media, such as the home video, have been shown to have profound social consequences. More recently, the Zapatista rebels in Mexico have exploited the Internet in what has been characterized as a campaign of information warfare against an oppressive government and international economic system. Mark Poster (1995) has emphasized this emancipatory potential of new information technologies such as the Internet. On the one hand the Internet has helped accelerate linguistic standardization and cultural homogenization, but on the other hand, 'the Internet seems to encourage the proliferation of stories, local narratives without any totalizing gestures and it places senders and addressees in symmetrical relations' (Poster 1995: 92). This is precisely what the pioneers of information warfare have shown us; the limits of control appear as people take personal control of media and recording technologies outside the global monopolies.

## *Destroying the word virus*

> Europe owes much of its murderous history to errors of thinking engendered by the alphabet.
>
> (Gray 2003: 58)

Orwell proposed a very simple solution to the political distortion of language – avoid the use of jargon oneself, and challenge jargon when hearing it from others (Orwell 1994). He provided five simple steps to help accomplish this. Never use a metaphor which one is used to seeing in print, i.e. don't become a slave to fashionable terminology. Never use a long word when a short word will do the job just as well, don't waste words. Never use the passive tense when the active can be used. Don't use jargon if an everyday equivalent is available. Finally, one should break any of the previous rules rather than writing something that is poorly expressed or barbarous.

Jargon is not the only obstacle to be overcome when considering language as part of a cybernetic system of control. In his book *Straw Dogs* John Gray (2003) argued that if we are to adequately address the violent legacy of human history then we must somehow face up to problems inherent in the metaphysics of language. In the ancient world, the invention of the alphabet allowed the development of new philosophies, which created a brave new world of abstract entities and transcendental forms. This world of transcendental forms, included ideas such as the Good and the Beautiful, which were believed to be universal forms that were more truthful even than the reality of the senses. This has been the basis of much of Western philosophy since Plato. This belief in the eternal and unchanging nature of these ideas led Nietzsche to remark in *Twilight of the Idols* that, 'I fear we are not getting rid of God because we still believe in grammar...' (Nietzsche 1990a: 48). Plato's system has been roundly criticized for its belief in a world comprised of words and concepts that were supposedly superior to the world of the senses (Deleuze 1990). According to Plato, human justice and human beauty were merely a simulacra, or a 'bad copy' of the original ideal forms. Plato argued that one could never have justice in this life, but only a poor imitation of true Justice. Life itself was judged by him to be a poor imitation of a transcendental form. According to Gray (2003), the belief in a transcendental reality has supported the folly that humanity is somehow outside of nature, striding above the rest of creation with its access to these higher truths. This foolishness has also meant that the human race is alone amongst the animals in killing and fighting wars in service of its metaphysical abstractions.

Metaphysical preconceptions are built into the grammar of language, specifically relating to the definite article 'THE' and the verb 'IS'. (Burroughs and Odier 1989: 201–202) claimed that 'The IS of identity is in point of fact the virus mechanism'. The verb 'to be' plays a very peculiar role in language because it posits 'being' as an activity that is performed by a presupposed subject – 'I am'. However, this subject is nothing other than the virus; 'I' is a parasite that endures as a word virus subsisting on its human host. Burroughs explained this in the following terms, 'The word BE in English contains, as a virus contains, its precoded message of damage, the categorical imperative of permanent condition. To be a body, to be nothing else, to stay a body' (Burroughs and Odier 1989: 200). Language is a customs officer asking for credentials. The human is trapped by the word virus, and is forced by this imperative of language to identify itself as a word. In contrast to this, Burroughs argued that,

> What ever you may be, you are not the verbal labels in your passport any more than you are the word 'self'. So you must be prepared to prove at all times that you are what you are not. Much of the force of the reactive mind also depends on the falsification inherent in the categorical definite article THE.
>
> (Burroughs and Odier 1989: 201).

The definite article 'The' and the verb 'to be' privilege the idea of stasis and permanence over transformation and becoming, they are both residues of a primitive egoistic psychology buried deep in the subsoil of language. Burroughs proposed a move away from the language of authority and identification, which he called 'family talk, mother talk, father talk, cop talk, priest talk'. Instead he proposed a silent hieroglyphic language, or an alphabetic language devoid of the definite article. Such a language allows a description of events as transformations, moving in between states without demanding that one be stamped with an authorized identity. Burroughs proposed to banish the word virus completely from the human organism. He scorned the kind of pernicious moralism that is frequently found in the output of the mass media. The 'RIGHT virus' is perhaps the most unpleasant and virulent form of word virus:

> . . . my contention is that evil is quite literally a virus parasite occupying a certain brain area which we may term the RIGHT center. The mark of a basic shit is that he has to be right . . . This RIGHT virus has been around for a long time, and perhaps its most devoted ally has been the Christian Church: from the Inquisition to the Conquistadores, from the American Indian Wars to Hiroshima, they are RIGHT RIGHT RIGHT.
>
> (Burroughs 1986: 16)

Grammar brings with it a whole set of metaphysical and psychological prejudices. For example, grammar tells us that a subject exists in distinction from its actions. This grammatical construction is related to a moral system, whereby a doer can be held responsible for their actions. Nietzsche (1990a) pointed out that this is not an empirical claim but is an entirely metaphysical presupposition. Language allows us to talk about the agent of an activity, where only the effect of that activity can be detected empirically. For example, when we say that, 'the lightning strikes', this is simply the discharge of electrical energy, a pure activity, but language has attributed a subject to that activity that exists prior to the activity, i.e. the 'lightning'. According to Nietzsche's genealogical studies, the presupposition that a subject exists as the cause of every action derives from a primitive stage in the history of human psychology. His reasoning follows from the observation that language:

> sees everywhere deed and doer . . . which projects its belief in the ego-substance onto all things – only thus does it create the concept of the 'thing' . . . Being is everywhere foisted on, as cause; it is only from the conception of 'ego' that there follows derivatively, the concept of 'being' . . .
>
> (Nietzsche 1990a: 48)

We see something like this even in the most rational of philosophers, such as Kant, who posited the idea that man, as a rational being, lies outside the causal chain as a noumenal first cause. In Nietzsche's terms, if the ego is presupposed as a causal agent then, God can be defined as the biggest causal agent, the biggest 'thing', and one might add the biggest ego – 'I AM THAT I AM' (Exodus 3:14). This primitive psychology denies processes of becoming and difference, and in place of this holds firm a belief in eternal being, that has been defined in terms of identity, logic and duration.

There appear to be a range of strategies for destroying the word virus, along with the control that this exerts on our consciousness. The genealogical studies undertaken by Nietzsche, Foucault and Deleuze have begun to raise our awareness of the presuppositions of language and the prejudices that they may engender. Foucault's later work has conceived of a self that can be developed independently from the normalizing institutions and the diagnoses offered by so-called experts of the self. All of these thinkers have proposed a movement away from a philosophy of being to one that focuses on becoming and difference. On the simplest level, one could follow Orwell's advice that propaganda can be countered by avoiding the use of fashionable metaphors and jargon of any form. Alternatively, Deleuze (1995) has suggested that we allow for silence and that we should make the most of gaps that exist within our systems of communication. Information warfare can be seen as a general strategy for achieving such aims. The success of groups such as the hackers and the Zapatista rebels is in part due to their capacity to disrupt the mass media and make use of the gaps they create to protect their identities and further their own communities' purposes. Burroughs has suggested the use of cut-ups and has even experimented in hieroglyphics to silence the incessant chattering of the word virus. Marketing executives and politicians have always been experts in the use and abuse of language and have turned it to their own purposes as part of a system of propaganda and control. Nonetheless, language can always be re-appropriated, whether it is to ridicule the jargon of management, or in the private language shared between friends, or the messages that have been sent out by Subcomandante Marcos from the Lacondon jungle proclaiming that 'Our Word is our Weapon' (Marcos 2001).

## Reconsidering metaphors of organization

This chapter has discussed the importance of the virus as a metaphor for understanding organizations and networks of communication today. The virus serves as a useful metaphor for understanding organizations on many different levels. First, organizations are designed to some extent to ward off the damaging effects of viral contamination. Historically, Foucault has traced the origins of disciplinary organizations to the plague cities of the seventeenth century. Today, organizations are inventing new methods of policing and social control to deal with a network society, which has become a society of universal contagion. Viruses are manifest as technological infections of machines, as biological infections such as AIDS and SARS, and as cultural infections of language and the word virus. The virus is a disruptive force. It can be understood as a metaphor for the troublesome side of the network society, where it is the prime mechanism for the disruption of these networks. In its incarnation as the 'ideavirus' and word virus, the metaphor also works in a productive sense, illustrating the great power of networks of communication. Ideas and even language circulate in a viral manner, spreading rapidly via communication nodes and ports around the globe. Historically, there are many parallels between the spread of viruses and the spread language. Identity itself has been conceived as a virus in that it is a parasitic organism subsisting on the human body with no other function than to reproduce itself.

There has been scant attention paid to such considerations within the discourse of organization studies and management. Gareth Morgan's best selling book *Images of Organization* has outlined a series of metaphors for understanding modern organizations, but neither the metaphor of the network nor the virus are numbered among them. Morgan's book describes the concept of organization variously as a machine, an organism, a culture, a political system, a brain, a psychic prison, flux and transformation, and a system of domination. The emphasis on the use of metaphor in understanding organization does serve to highlight its linguistic features and downplays the idea that an organization is a 'thing' that exists independently from one's perceptions and the metaphors used to describe it. One must be careful to make certain that such metaphors do not become fixed stereotypes that serve as a shortcut for genuine creative thinking about the problems and issues in which one is embroiled. Morgan's work gets close to a viral understanding of organization in his discussion of the 'flux and transformation' metaphor. He explains that, 'Egocentric organizations draw boundaries around narrow definitions of themselves and attempt to advance the self-interest of this narrow domain' (Morgan 1997: 260). This argument undermines the view that organizations exist as 'things' that are somehow separate from their environment. Morgan uses this metaphor to criticize the kind of instrumental thinking that has caused massive environmental destruction over recent decades. In this respect, he comes very close to appreciating the ego or 'I' as a parasitic entity, in much the same way as William Burroughs. The body of work associated with Actor Network Theory also provides an interesting avenue for exploration, which understands organizations as heterogeneous assemblages of people, ideas and things and presents a challenge to the notion that organizations are unitary, bounded objects. Gibson Burrell's book, *Pandemonium*, provides one of the few sustained accounts of the concept of the unitary organization and challenges the traditional idea of an organization as a thing. Burrell is one of the few writers in organization theory who has taken the work of Foucault and

Nietzsche seriously, highlighting the processes of disease, fear, pain, loathing, identity and power in the formation of organization, in what he terms a 'retro-organization theory'. The idea of a rational, conscious actor has only minor explanatory power in such a theory, which highlights instead the significance of the unconscious forces at play in the process of organization. The virus is the perfect metaphor for understanding the network society and the organizations within it. It describes the spread of ideas within communication networks, but it also characterizes the main method of disruption within these very same networks.

# Part III

# Changing power relations

# 6 Power relations in the Information Age

He who knows everything fears nothing.

(Joseph Goebbels)

We're moving toward control societies that no longer operate by confining people but through continuous control and instant communication.

(Deleuze 1995: 174)

Imagine a society where one's employers could dictate who your friends were or suitable lovers. This is precisely what happened to Virginia Rulon-Miller, an employee of IBM, who was sacked on the grounds that she might leak valuable corporate information to IBM's competitors. The evidence against her was that she had been dating a man who worked for a competing firm. She took IBM to court for unfair dismissal where the law found in her favour. Nonetheless, this case raises serious issues about the extent of corporate power over employees' private lives, and the complexities that may be involved in regulating and containing valuable corporate information. The integrity and security of information is gaining in prominence as our economies increasingly value information and develop new techniques for knowledge management, corporate intelligence, and the protection of intellectual property. The systems of banking and finance upon which many of the world's economies are based have come under new threats with the rise of crime such as identity theft. This kind of fraud has been facilitated by the use of on-line banking, where fake websites and other techniques have been constructed to steal the personal details of a bank's customers (Denning 1999, Mitnick 2002). Banks may be reluctant to publicize such problems for fear that they may deter customers from using their services. Identity theft may also be difficult to deal with because it can be done remotely, from anywhere within the communications network. The Internet is powerful insofar as it facilitates real-time communication on a scale never seen before, but it is vulnerable for precisely the same reason; it can be attacked from any point in the network.

We appreciate the effects of power because they become apparent in the struggles that surround us. Information warfare can be explained, at least to some extent, in terms of the exercise of power. Different techniques of power can be employed depending on the information resources available to the individuals or groups concerned. This chapter will examine how power relations in society are changing with the advent of the information-age. Information societies are moving from disciplinary methods of social control to non-disciplinary methods of control that focus on regulating flows of information rather than on more direct control of the body. The information society is distinguished by a number of social and technical innovations and associated institutions. Perhaps the most profound social innovation has been the invention of the concept of intellectual property rights and the increasing treatment of information as a commodity. The technological advancements in personal computing and the inception of the World Wide Web have allowed for further social developments, such as on-line communities and e-commerce. The language of commerce has spread in step with the growth of global networks of communication and the downfall of the Soviet system.

On the one hand, the information-age has seen an unparalleled rise in the access to information and communications technology, through television, the Internet and telephone technology. On the other hand, new institutions are now being developed for the regulation of access rights to this information and the protection of its integrity. The software giant Microsoft frequently flexes its corporate muscle to protect its information resources, notably through the law courts when suing competitors for possible infringements of its intellectual property rights, or by using its Public Relations department to discredit competing organizations, so undermining the public's perception of the latter's products. A particularly vicious campaign has been waged against open source software such as Linux, where each organization has used the information resources available to it in order to undermine the other.

Where intellectual output can be a more valuable commodity than material goods, techniques must be tailor-made for its control and regulation. The media today is full of examples of new technologies of control: a parent installs an electronic tag in their 11 year old child as a safety precaution; a group of biotech companies deny the licence to cheap pharmaceutical drugs to desperate third-world countries; a woman is sacked by IBM for dating a man who works for a competing organization. What is clear from all these cases is that the apparatus of social control is changing and that many of these changes are arising from the use of new information technologies and the need to protect the integrity of these systems. Michel Foucault coined the term 'biopower' to describe the various techniques that began to emerge in the nineteenth century to organize and control the collective energies of the masses within the growing towns and cities. The concept of biopower has been taken up and developed by other important thinkers, including Gilles Deleuze (1995) and Hardt and Negri (2000), to describe innovations in social control during the late twentieth century. Hardt and Negri have reformulated the concept of biopower to account for innovations in the productive apparatus over the last century, specifically the rise of global networks of communication and the increasing importance of 'immaterial labour' to the means of production. The following argument will analyse these developments in terms of their underlying relations of power. We will begin with a brief look at the concept of power, drawing on Foucault's historical studies, and then develop these ideas with respect to changes associated with the network society.

## What is power?

> There is no power without potential refusal or revolt.
>
> (Foucault 2000b: 324)

Looking at the *Concise Oxford English Dictionary* for a simple definition of power, one is faced with no less than 18 alternative definitions, many of which are broken down yet further. Power is by no means a straightforward or simple concept. The approach to power taken in this chapter derives from the work of Michel Foucault and his conception of power relations. Foucault was himself unsure whether one could describe the essence of power, or say what power actually consisted of, but one could describe the many ways in which it is exercised. Foucault argued that viewing power simply in terms of violence and coercion covers only a limited range of behaviours and was a very impoverished and restricted conception of power. In contrast to this naïve definition, a power relation is often far subtler and more sophisticated than naked violence. Foucault outlined a history of techniques for the exercise of power, and explored the social relations that these techniques bring into play. This focus on techniques of power also challenged the received wisdom, which saw power as a characteristic springing from individual persons. In fact, those in authority may be just as much the victims of their institutional position as those lower in the hierarchy.

Power defines the relations between different people before it can be said to pass between them. If a person can be said to possess power, which they exercise over others, then one should also recognize that this power relation circumscribes their respective roles and the possibilities that such roles afford. This is how one might best understand the relationship between judge and criminal, doctor and patient, parent and child, priest and confessor. The way in which one understands oneself and others, one's very identity, is entwined in these power relations. Power should not be understood as a primarily repressive force because, 'power produces, it produces reality; it produces domains of objects and rituals of truth. The individual and the knowledge that may be gained by him belong to this production' (Foucault 1977: 194). Better terms to understand the nature of power may be indicated in the following: to incite, to induce, to seduce, to enlarge, to limit, to make more probable (Deleuze 1988). Power is often invited into a situation rather than being an unsolicited imposition from the outside.

For Foucault, power was an unstable concept, continuously mutating its form and evolving new techniques for its application. He traced at least three different stages in the evolution of power: sovereign power, disciplinary power and a more abstract form of biopower. Sovereign power is defined by the spectacle of punishment, where the criminal is subjected to horrifying tortures in full view of the public. The punishment is testament to the dreadful power of the king, who had the god-given right to grant life or death to his subjects. Foucault described sovereign power as a 'society of blood' (Foucault 1981: 147), concerned with lineage, caste and similar feudal distinctions. The punitive apparatus was both extreme and marginal, and institutions like the prison played no great role in this society. If these power relations were to be described in topological terms, then power would appear to radiate downwards onto the rest of society from a higher god-appointed position, as shown in Figure 6.1.

With uprisings against the traditional monarchic institutions in the seventeenth and eighteenth centuries, the grounds for sovereign power became severely undermined. Admittedly, some kings were deposed only to be replaced by yet others, and experiments on the monarchic theme reappeared under the guise of England's Lord Protector and France's Emperor Napoleon. Nevertheless, the damage done to the institution of the monarchy was permanent. During this period of revolution, new techniques for the administration and control of social life were developing in response to catastrophic events, such as civil war and the plague. In *Discipline and Punish*, Foucault described a historical movement between two forms of disciplinary power, the first being the 'discipline-blockade' which developed from the quarantine of cities during the plague years. As the plague spread through the cities of Europe, citizens were confined to their homes and their movements were closely supervised. This form of power was introduced to deal with special circumstances, and played a relatively marginal role in the constitution of the social order. The second form of

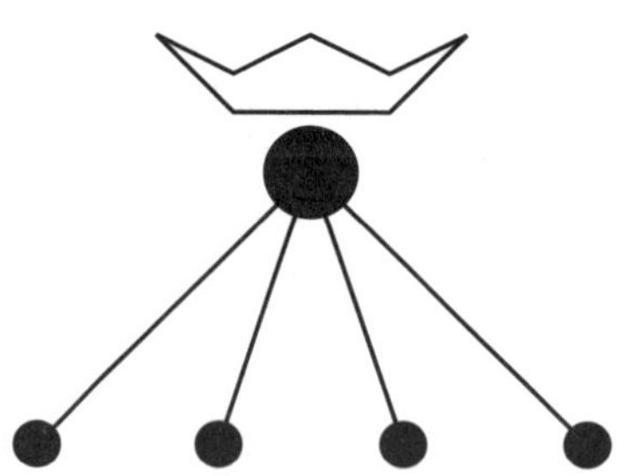

*Figure 6.1* The sovereign diagram.

discipline identified by Foucault was the 'discipline-mechanism' which was a more abstract form of power that emerged in institutions across the social spectrum including, schools, factories, hospitals and barracks. The discipline-mechanism found its ideal expression in the Panopticon, which was an architectural design conceived by Jeremy Bentham as the perfect prison. It was designed as a circular building with a central surveillance tower encircled by a wall of prison cells. Every cell in the Panopticon could be seen from the central tower, and was lit from behind to make it clearly visible to the guards. However, the cells themselves were cut-off from their neighbours to prevent any lateral communications between prisoners, who were treated as the passive objects of any communication. Since they could not see into the central tower, they could never tell when they were being watched, leading to a feeling of continuous surveillance. Ideally, the inmate would internalize this state of continuous surveillance to such an extent that they would become the guardian of their own actions, and the presence of an external authority could be dispensed with.

The prison moved from being on the periphery of society to being one of its central organizing mechanisms. Disciplinary power was by no means restricted to the prison, and Foucault himself described a 'swarming of disciplinary mechanisms' throughout all kinds of social organizations (Foucault 1977: 211). That is what he meant when he wrote that power struggles tend to be transversal, where similar struggles crop up in organizations with very different aims; discipline became a way of seeing problems in schools, barracks, factories and all manner of social institutions. One could spend one's entire life within the walls of a disciplinary institution, advancing from the school to the factory, and finally into the hospital. Punishment moved from being a spectacle and a manifestation of the absolute power of the king, to being a relatively modest means of discipline and reform. Foucault outlined a number of methodological guidelines in his essays on power, which I shall attempt to follow here. One important point he makes is that there is no power without the possibility of resistance, and consequently, one can trace mutations in the exercise of power by following the points of resistance to power. The struggles discussed below and elsewhere in this book revolve around issues of property rights and a broader concern about access to knowledge and information. Such struggles can offer a guide to the empirical reality of power, where innovations in the exercise of power follow in their wake.

## The diagram of power

> Visibility is a trap.
>
> (Foucault 1977: 200)

Foucault's work examined the historical emergence of a vast number of techniques for the exercise of power, including torture, dressage, confinement, the timetable, the examination, hierarchical observation, and normalizing judgement. However, these techniques take on a different role and prominence depending upon the underlying apparatus of power of which they are a part. This underlying apparatus of power is termed by Foucault the 'diagram of power'. He explained the concept of the diagram in the following paragraph:

> The Panopticon must not be understood as a dream building; it is the diagram of a mechanism of power reduced to its ideal form; its functioning, abstracted from any obstacle, resistance or friction, must be represented as a pure architectural and optical system; it is in fact a figure of political technology that may and must be detached from any specific use.
>
> (Foucault 1977: 205)

The architecture of surveillance, which underpinned the Panoptic ideal, became the pervasive form of social organization within disciplinary institutions. The Panopticon was not simply one control mechanism amongst others, because it served to intensify and integrate a series of diverse mechanisms of control, such as confinement, surveillance, the timetable, the examination and dressage. Disciplinary methods were in existence when sovereign power held sway, but only as a bundle of disparate techniques. Foucault (1977: 224) argued that with the Panopticon these techniques crossed a technological threshold to form the disciplinary diagram of power. The diagram of power reveals a particular relation between forces and the way in which time and space are distributed according to these forces. If these power relations were to be described in topological terms they would be represented as one central node surrounded by a circle of subordinate nodes, as shown in Figure 6.2.

The diagram of power acquired a predominantly visual emphasis in the

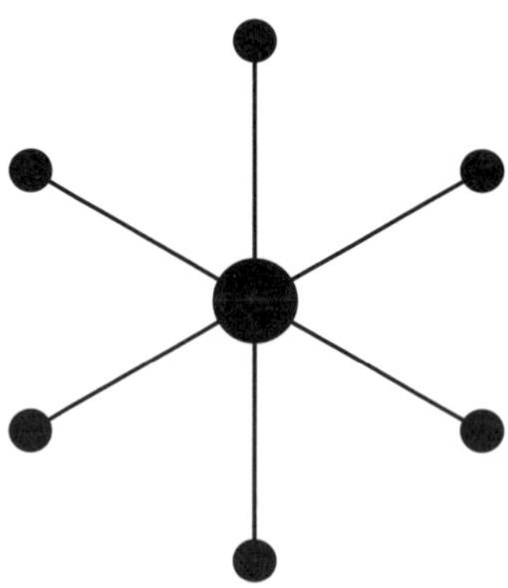

*Figure 6.2* The Panoptic diagram.

Panopticon, however, this need not always be the case. The diagram of power describes a relation between forces, and should not be confused with a purely visual architecture. Deleuze (1988) has explained that as a relation between forces, the diagram of power operates simultaneously on a discursive level and a non-discursive level (visible, institutional). The conditions that allow one to say a certain thing are not the same as the conditions that allow one to see a thing in a certain way, 'There is a disjunction between speaking and seeing, between the visible and the articulable: 'what we see never lies in what we say' and vice versa' (Deleuze 1988: 64). Where the prisoner is the visible subject of the prison, the delinquent is the discursive subject of penal discourse. The discourse on delinquency can evolve somewhat independently from the prison, although there is a clear relationship between the two. The diagram of power acts as an abstract machine that describes the relations between seeing and speaking. It brings together forms of visibility such as the prison architecture with forms of expression such as the legal and psychiatric discourse on delinquency. The fact that power relations combine forms of visibility with forms of discourse will be investigated later in this chapter in greater detail, with a particular focus on the technological advances in communication and surveillance that have emerged over the course of the twentieth century.

## From discipline to biopower

The section of *Discipline and Punish* that deals with the Panopticon also marks the increasing importance of the police. The police are an important arm of the disciplinary society but they are not themselves tied to any enclosed institution of discipline. The remit of the police is far broader, and they operate by 'disciplining the non-disciplinary spaces' (Foucault 1977: 215). The police apparatus is co-extensive with the state itself. In a later essay, Foucault observed that the object of the police is the entirety of social existence, 'The police includes everything' (Foucault 2000b: 319). In the terms of the disciplinary society, the police ensure 'an ordered maximization of collective and individual forces' on behalf of the state' (Foucault 1981: 24–25). In the police we see one of the major instruments of biopower, the prime means for the control of populations. The concept of the police is not restricted to the supervision of criminality and political revolt, but has more general implications for what Foucault termed 'biopolitics'. This concept is grounded in the fact that there is a continuum between the government's general concern for social welfare (i.e. policy) and the more specific tasks of policing deviance. Policing the population is not restricted to a dedicated police force but is a function of all branches of the welfare state concerned with 'biological process: propagation, births and mortality, the level of health, life expectancy and longevity...' (Foucault 1981: 139).

The term 'biopower' was coined by Michel Foucault to describe 'what brought life and its mechanisms into the realm of explicit calculations and made knowledge-power an agent of transformation of human life' (Foucault 1981: 143). He traced the history of biopower on a number of different levels, all of which are directed at the treatment of the mind and body in some respect. Foucault stated that the era of biopower began with an explosion of techniques for 'the subjugation of bodies and the control of populations (Foucault 1981: 140). In early capitalism, biopower concerned the transformation of the body to make it useful and productive. At the level of government, biopower also concerned the administration and policing of populations. The techniques of biopower have been taken up and elaborated by several professional associations, such as the medical and legal professions, and explain the latter's prominence and status within the social hierarchy.

Hardt and Negri (2000: 324) have observed that NGOs, such as Amnesty International and the Red Cross, serve a biopolitical function because they meet the needs of life itself. These organizations are directly concerned with the moral and legal issues of biopower. For example, the human rights organization Liberty has recently objected to moves within the British civil service to combine and centralize the databases from its different branches of government, including taxation, health and crime. Liberty has argued that such integration, whilst useful for administrative purposes, could threaten the individual's right to privacy. During the twentieth century, biopolitics has also moved into the realm of private industry with the emergence of huge corporations to deal with the provision of health insurance, life insurance and pensions. Biopolitical information is no longer the preserve of the state but is accumulated on behalf of private enterprises by massive databases, which collect all sorts of data for the purposes of marketing and consumer profiling (Elmer 1999). In many respects the literature on the practice of business administration and management is directly concerned with developments in biopower. Frederick Taylor's pioneering work on 'scientific management' and the extraction of the maximum effort from the labouring body is very much in keeping with the disciplinary techniques described by Foucault. Since Taylor, management thinking has developed far more abstract methods for the administration and control of the body such as ergonomics, accountancy and human resource management (Townley 1994, Jermier *et al.* 1994, Mckinlay and Starkey 1998). These developments have begun to focus on governing the contents of the employee's mind with the use of knowledge management programmes. The implications of these developments for the concept of biopower will be discussed in more detail in the following sections on immaterial labour and networks of communication.

## Communication networks and immaterial labour

> Those who have been thoroughly informed lend themselves to thorough utilization.
>
> (Adorno 1991: 72)

Disciplinary power emerged at around the same time that capitalism became a dominant force throughout the world. The pages of *Discipline and Punish* are full of allusions to Marx's *Capital*, although the latter is rarely referred to explicitly. Foucault himself remarked that mutations in the economic infrastructure went hand in hand changes in disciplinary technology, '...it would not have been possible to solve the problem of the accumulation of men without the growth of an apparatus of production capable of sustaining them and using them; conversely, the techniques that made the cumulative multiplicity of men useful accelerated the accumulation of capital' (Foucault 1977: 221).

Foucault's work shows us that power relations are constantly mutating, depending upon human inventiveness and historical contingency. Power relations change as they react to and incorporate novel forms of resistance. The nineteenth century witnessed new struggles rising up around the new modes of economic production and exploitation. A new political class of labourers and machine operators came into being with the rise of the 'working class'. In contrast, Foucault observed that today's struggles are more concerned with forms of subjection and subjectivity (Foucault 1982: 331–332). Hence, his work was much more concerned with the treatment of subjectivity in terms of criminality, madness and sexuality, and his later works cover normalization and technologies of the self. This relationship between power and the subject is crucial to the emerging forms of immaterial labour that require intensified levels of cooperation.

Hardt and Negri (2000) have distinguished between three distinctive modes of production: (i) agriculture and the extraction of raw materials, (ii) industry and the manufacture of durable goods, and finally, (iii) service provision and the manipulation of information. Economic activity is increasingly characterized by the third of these modes of production. It is important to note that when production moved from being largely agricultural to largely industrial, the remaining agricultural production was itself reorganized along industrial principles. Exactly the same process of reorganization is taking place today with respect to the manufacturing industry, not only is it supported by more information, it is itself being transformed into another kind of service: 'all production tends toward the production of services, towards becoming informationalized' (Hardt and Negri 2000: 286). The factory has been moved into the home, and the living room, not simply in terms of teleworking, but in terms of consumption, marketing and advertising, the development of social skills, and the training of the intellect and emotions. Hardt and Negri describe this situation in the following terms: 'social relations completely invest the relations of production, making impossible any externality between social production and economic production' (Hardt and Negri 2000: 209).

During the 1980s and 1990s, the world's richest countries underwent a massive structural change towards a more service-orientated economy. This restructuring took place around the world following two broadly different approaches. The approach taken by Germany and Japan was to adopt information technology to informationalize and support their existing manufacturing bases, whereas the approach pioneered by the USA and the UK led to very high levels of unemployment in their manufacturing industries and a far more dramatic shift in employment towards financial, informational and leisure services. Currently, over 80% of employment in the US economy is in the service sector, and in the UK it is slightly lower at about 73%.

As noted above, less-developed countries lack the necessary educational and technological infrastructure to make this move. The product of work in a service industry is often not a physical commodity, but an emotion (e.g. comfort, pleasure, reliability) or the provision of information (e.g. consultancy, training). This kind of work can be called 'immaterial labour', since although it still requires labour its product is not a durable good (Hardt and Negri 2000). Immaterial labour is made of three broad areas: communicative labour in informational networks, interactive labour in symbolic analysis and problem solving, and affective labour in the provision of a service. These kinds of jobs require what might be termed 'social skills' and a high quality of communication and cooperation. In this work, there is a greater focus on concepts such as education, which is now understood in terms of life-long learning and increasing flexibility. The flexible worker is on-call at a moment's notice and is able to retrain and develop new skills to suit the demands of the job at hand. Concepts such as real-time control and teleworking have led to the blurring of boundaries between office/home, work/leisure, and other traditional institutional boundaries. Organization theorists are now discussing the emergence of a 'flexible' workforce that can adapt to the turbulent environment created by the new networks of communication and production (Boje *et al.* 1996). The focus on 'flexibility' demands characteristics such as mobility, time flexibility, educational qualifications and communication skills (Krings 2003). The creation of a flexible workforce has also seen a rise in the number of part-time jobs and work contracts that allow the conditions of employment to be changed at short notice. Some critics have pointed out that much of this flexibility is nothing more than an outright attack on working conditions and the rights of employees. In her evaluation of current European Union policy on social exclusion, Ursula Huws (2003) has observed that forms of social exclusion, such as unemployment and poor education, have been a key driving force in the creation of a flexible workforce rather than an unintended consequence.

The division between labour and other social activities is becoming increasingly blurred. For example, Hancock and Tyler's (2000) study of the work of flight attendants has highlighted the diverse aspects of an attendant's labour that have been subjected to scrutiny and regulation, including amongst other things: the way they look (height, weight, shape, attractiveness, uniform, hairstyle, hair colour, make up, skin, teeth, smile), the way they sound (tone of voice, accent, volume), and the way they move (poise, posture). The flight attendant must look natural whatever the circumstance and may take great pains to appear so. One of many strange incidents that Hancock and Tyler bring to our attention is the case of an attendant who was asked to dye their hair in order for it too look 'more natural'. Her natural colour apparently was not 'natural' enough to fit the corporate aesthetic. These aesthetic qualities of a flight attendant's behaviour are considered to be a part of the service provided, and are emblematic of the corporate image. The aesthetic qualities of the attendants is no less a corporate resource than the physical power of their fleet's engines. Much of this aesthetic labour is done outside of working hours and yet is not explicitly rewarded as such; it forms part of what Hancock and Tyler refer to as the invisible labour process. For example, a strong regional accent may require moderation before it fits the company image, and this may require a great deal of practice before it is an acceptable accent for work. Any excess weight an attendant puts on whilst at home or on holiday, cannot be magically shed on turning up to the office, and constant self-discipline is required over the attendant's behaviour and overall appearance. This is no less true for a flight attendant than for a management consultant, or any other worker in the service sector. Recent research on behalf of the European Commission has found that even in the IT industry employers tend to value social competencies, such as the ability to cooperate, to consult and to communicate, even more highly than they value purely technical skills (Moniz 2003). These social skills are developed outside of the workplace, as part of growing up in a family, as part of a class of students at school; in fact these skills are developed during almost any communal practice. A new level of cooperation is essential with the emergence of immaterial labour, where the production process directly concerns the production of social relations, and cooperation is at once a product of these networks and a prerequisite for their formation.

Hardt and Negri explain that affective labour and information-based forms of production are comparable because both are immaterial products and both produce '. . . social networks, forms of community, biopower' (Hardt and Negri 2000: 293). They draw on Foucault's concept of biopower, which was originally developed for the social practices of eighteenth and nineteenth century capitalism, and rework this concept for the economic and political practices of post-industrial capitalism. One of the central differences with Foucault's work in their reformulation of biopower concerns the increasing importance of immaterial labour, and how control is exercised over bodies and minds in order to extract value from this form of labour. This argument also builds upon Deleuze's (1995) short essays on the 'control society'. The idea behind the control society is that the twentieth century has witnessed a movement away from disciplinary societies to more invasive societies of control. This does not mean that disciplinary institutions have disappeared, but that their authority is no longer confined to particular institutions. Instead, power has become increasingly integrated into social life by way of interconnected networks of communication and information.

## Controlling flows: connecting nodes

There are significant differences between the techniques of power associated with the Panopticon of the eighteenth century and techniques of control today. The node in a network is very different from the cell in a prison. Rather than preventing lateral communications, the whole purpose of the network is to proliferate such links. These nodes include the rail terminals, airports and seaports that make up the world's transportation networks, and they also include the computer systems that form the local and wide area networks that make up the Internet. Cities may also be seen as primary nodes in the global circulation of information and commodities. Paul Virilio has described the emergence of a super city, which he calls the 'omnopolis', essentially the result of the hundreds of millions of users who can participate in some way in the virtual world of the Internet. Concentrations of virtual nodes tend to parallel pre-existing cities, such as Manhattan, London, Tokyo, Toronto, Munich and Stockholm. London alone has more Internet domains than the whole of Africa (Castells 2001). Fuchs (2003) has noted that for all the utopian talk of a virtual workplace, companies in the high-tech multimedia sector still tend to congregate in close physical proximity, hence the phenomenon of Silicon Valley. Fuchs' research has found that tacit knowledge plays a very important role in generating innovation, and such tacit knowledge requires physical proximity and the development of a close community for its exchange. Another defining feature of these cities is that they tend to be characterized by the close proximity of extremely unequal populations (Hardt and Negri 2000: 336). The division between first world and third world means little in cities, such as Los Angeles and San Paolo, where shanty towns and ghettos have grown up around their productive centres.

The kind of surveillance in operation within these networks is of a different kind from that outlined by Foucault in his work on the prison society. The primary object of surveillance in the Panopticon is the body, but this is not the case with many modern forms of surveillance, which operate largely on abstract data. McKinlay and Starkey (1998) have traced the rise of the accounting function in companies such as General Motors and Ford as a form of 'corporate Panopticon'. The object of this form of surveillance is not the human body, but the various costs and revenues generated by particular managers and their departments. The behaviour of managers is made visible through the use of numbers, where a variety of accounting techniques are employed to measure their relative performance. Mark Poster's (1995) work has highlighted the increasing use of databases to monitor and control behaviour, whether it be targeted at consumers or criminals. This work has described innovations in surveillance where the subject participates voluntarily. Participation can be encouraged by offering a small reward in exchange for data about everyday transactions, for example, when credit card companies offer increased credit ratings for increased use of their cards, or when retail outlets offer 'points' to customers who use their loyalty cards. Zuboff's seminal study of the computerization of workplace communication described the emergence of the 'information Panopticon', which is used to record and monitor exchanges between employees. Zuboff (1988) also observed a marked reluctance on behalf of employees to use such media where their communications may be recorded. The success of these electronic forms of surveillance can be explained very simply because they are more efficient than the strictly Panoptic approach.

Surveillance is essential for the maintenance of the integrity of communication networks. Over recent years, companies have been increasingly subject to attacks on their information resources. They may be subjected to unauthorized access and the theft of vital information, or they may suffer a direct attack against their communications networks by computer viruses. Kevin Mitnick (2002) has outlined a myriad of ways in which a person can gain access to valuable corporate information by using their social skills without needing a high level technological expertise. As previously mentioned, Mitnick uses the term 'social engineering' to distinguish the use of psychological manipulation from the purely technical skills of the hacker. The very same characteristics that are essential to immaterial labour, such as trust and cooperation, are exactly what the social engineers exploit to gain access to restricted information. They are able to exploit a person's willingness to help. The social engineer tends to avoid direct personal contact, preferring a telephone call or email to establish a relationship, and attempts to persuade the person on the other end of the phone to hand over the necessary access codes or other relevant information. They can do this in a variety of ways, which generally involve posing as a reliable authority, for example by using local knowledge of the target organization, knowing the corporate jargon and knowing which names to drop. Mitnick describes this as 'speakeasy security', which 'relies on knowing where desired information is, and using a word or name to gain access to that information or computer system' (Mitnick 2002: 80). Speakeasy security is part of the fabric of the informal organization, and may be difficult to bring under formal control. Knowledge of formal and informal codes or jargon, whether legitimate or otherwise, is a means of access to flows of information and wealth. The integrity and security of these flows depends upon close monitoring of people's access to them, and developing techniques by which this access can be restricted. The rules of access constitute strict hierarchies of information and wealth within the network society. Zuboff's own pioneering study of the use of electronic communications within organizations concluded that, 'As the electronic text become the symbolic surrogate for the organization's vital activities, access rules become the fundamental equivalent of organization structures' (Zuboff 1988: 356). The following sections will explore the extent to which the information age is characterized by new forms of social divisions and innovations in technologies of integration.

## The social divide

> As soon as one source of technological inequality seems to be diminishing, another one emerges...
>
> (Castells 2001: 256)

Many commentators on the information age have noted that many of the existing social divisions have been perpetuated in the divisions between the information rich and the information poor (Castells 2001, Webster 2003, Fuller 2001). These social divisions may be associated with poverty, and have also been analysed under the relatively new idea of 'social exclusion'. Whereas poverty is essentially defined in relation to basic needs, social exclusion is defined more broadly in terms of both cultural and material deprivation, depending on a person's access to a broad range of things, including information, education, wealth and employment. A recent research project supported by the European Commission found that employment patterns are clearly segregated in terms of both class and gender (Webster 2003). Women make up the majority of the workforce in the relatively low-skilled sections of the information society, such as in call centres or secretarial positions. Women are massively under-represented at higher levels of management, and tend to be paid less than male workers in comparable jobs. Webster points out that some service sector work may be seen as a natural 'feminine skill' that is not rewarded as genuine labour. Similar observations have been made by academic studies of other areas of the service sector, which have highlighted the existence of an invisible labour process associated with the emotional labour of the service, a form of labour that is not overtly rewarded (Hancock and Tyler 2000).

Manuel Castell's analysis of the network society shows that a digital divide exists along a number of different axes, based on gender, age, ethnicity, income and education (Castells 2001). In the US, in 2000, 70.1% of people earning more than 75,000 dollars had access to the Internet, whereas of those earning less than 15,000 dollars only 18.4% had access. People with a bachelor's degree were also much more likely to have access than those without. White Americans and Asian Americans were nearly twice as likely to have such access as African Americans or Hispanics. The gender gap was more marginal, with only a 2% difference in the proportion of men having Internet access as compared with women. Castells also noted that these gaps appear to be narrowing rapidly, at least in the US where this study was undertaken. The nature of access is also variable, with the elite being more likely to have higher speed access with wider bandwidth. Access to IT alone is not enough, and there are significant disparities in the level of education between different social groups, especially in terms of poverty and race. The world is divided into the information rich and the information poor, along exactly the same lines that characterize the divisions between the developed world and the third world. A large proportion of the world's population does not even have access to a telephone. Castells draws on the example of South Africa to show that poorer countries do not have the necessary skills and education to meet the growing demand in high technology sectors. In Russia, many private corporations, such as banks and multinational companies, have had to bypass their outdated state telecommunications networks, using their own private resources instead. At the moment around 87% of websites are written in English alone (Castells 2001). All of which quite clearly shows a marked division between the rich and poor in the new communications revolution.

Education is now acting as a means of social exclusion rather than as a principle of empowerment (Fuller 2001). Society is stratified along the lines of race and class, which is reflected in the underlying level of educational attainment within these strata. Fuller has argued that access to knowledge and the means of knowledge production are restricted by means of the control of expertise, credentials and intellectual property. Education is increasingly becoming a 'credentials mill', where it has become an essential key for accessing the other material benefits that society may have to offer. Again, the divisions between the developed and undeveloped world are reflected in their respective access to knowledge and good education. Fuller has also highlighted the potential problems surrounding the concept of bio-prospecting, where large Western biotechnology companies are buying up the patents on potentially valuable biological materials found in third-world countries. The conflicts arising from bio-prospecting are increasingly finding their way into the media, especially the operations of Monsanto, the world's largest producer of genetically modified seeds. Monsanto is currently being challenged by Greenpeace concerning its patent of a gene sequence of Nap Hal, a strain of wheat that is commonly used for bread making in India. Greenpeace argues that this is tantamount to thieving the work that has been done by the indigenous farmers who have cultivated this strain of wheat over many generations (*Guardian* 31 January 2004). It is quite possible that these patents will be used so that people who live in these areas may be charged for using such materials even though they may be naturally occurring in their environment. Another more pressing matter is the recent legal debacle in the pharmaceutical industry concerning the production of generic drugs for treating diseases such as AIDS. This disease is now the leading cause of death in Africa and yet governments and corporations have been embroiled in a controversial debate over how to deal with the situation (May 2001). Drugs for the treatment of AIDS have been manufactured by the Western pharmaceutical industry but were simply too expensive to be affordable to the millions dying of the disease in poverty-stricken areas. When some countries, such as Ghana and Brazil, tried to import cheap generic versions of these drugs from Indian pharmaceutical companies, legal action was immediately brought against them in the World Trade Organization on behalf of the US pharmaceutical industry for infringement of their intellectual copyright. The international outrage caused by this legal action eventually led the US companies to back down and now many of the drugs are being supplied at cost price (Stiglitz 2002). The debate surrounding the WTO's Agreement on Trade-Related Aspects of Intellectual Property (TRIPS) is ongoing and will continue to be a highly contentious issue. The WTO's own statement of the main issue is in terms of a balance between providing incentives for future innovations, whilst allowing people to use existing inventions in the short term. This fairly bland statement becomes far more contentious when put in the context of new drugs for the treatment of AIDS. In this context, it is a question of protecting either the rights of those who own the intellectual property, or the rights of those whose lives could be saved by the drugs (May 2001). Either way, it is clear that the divide between the information rich and the information poor is no more apparent than in the controversies that surround the protection of intellectual property rights. Now that we have explored the different forms of social division that have been emerging in the information society, we will examine advances in technology that are being used for social integration.

## Social integration and cybernetic modulation

> Only connect.
>
> (E.M. Forster)

The information age has seen the rise of new global institutions and frameworks for the transmission and regulation of information. Modern information technology has allowed a massive decentralization of information, which is dispersed throughout the architecture of the Internet. However, this has been accompanied by countervailing forces for standardization and re-centralization. Centralization is a means of concentrating power but it also serves the more pragmatic role of setting common standards that allow the whole network to function. At the present time the vast majority of Internet websites are in English only. This has already been discussed in the previous section as being symptomatic of a social, economic and technological divide, but it can also be examined as a mechanism for integration. An important prerequisite for the exchange of information is the sharing of common standards between the transmitters and receivers of information, and the global spread of English can be seen as one vehicle for achieving such standardization. Language is not the only means for standardization; for instance, in the eighteenth century the reformist philosopher Jeremy Bentham suggested that money be the measure of all things. More recently, Hardt and Negri (2000) have stated that money is a far more effective medium for overcoming traditional social boundaries and hierarchies than the medium of information technology alone. Capitalism itself operates as a massive network of commodities where every limit appears as a barrier to be overcome, whether it is social, political or technological in origin. The flow of money, information and people has become a planetary-wide mechanism for universal integration (Hardt and Negri 2000: 191).

The global flows of money and finance exercise huge power over the policies of both corporations and nations. The Nobel Prize winning economist, Joseph Stiglitz (2002) has argued that debt-ridden third-world countries are obliged to endure political and economic interference from global institutions such as the IMF and WTO in order to manage their debt payments. Gilles Deleuze once observed that money serves as a far more invidious means of control than harsher measures such as imprisonment, 'A man is no longer a man confined but a man in debt' (Deleuze 1995: 181). For the poor, debt has become a way of life, but even the middle classes are encouraged by the banking system to enter into debt as young as is lawfully permitted, by taking on housing mortgages and accumulating huge debts to pay for their university education.

In their book *Empire*, Hardt and Negri (2000: 345) describe three major sources of global control: the nuclear bomb and the ether are added to money as primary instruments of control. Together, these form an apparatus of control that operates beyond the traditional disciplinary institutions. As a threat to humanity itself, the development of nuclear weaponry led to the rise of world superpowers, which commanded a monopoly over such technology. The bomb has created an 'omni-crisis' and a global police force to deal with it, where the United States has emerged as the only nation with the resources to fill this role. This omni-crisis has taken on novel form with the development of the 'War on Terror'. Perhaps the major innovation of this new kind of war is that it is limitless both in time and in space. Undoubtedly terrorism will never end, thus the war is perpetual by its very nature. The second Gulf War was justified by the US and UK in terms of a UN resolution from 12 years prior, from the first Gulf

War. Is it any longer necessary for the ruling powers to declare war? In terms of space, the new war spans the globe. Terrorism can spring from any node within the global network. This war is characterized more in terms of a police action for maintaining global order, than as a war between different competing nations. (The reader is referred back to the discussion of Chapter Four where these issues are dealt with in greater detail.)

The third form of control that Hardt and Negri outline is 'ether', which is the medium of communication that is employed by the apparatus of command. Information technology allows control to be exercised through communication networks, remotely and continuously. Many of the latest innovations in management have exploited this new level of interactivity – such as Just-In-Time management, which exploits communication technology to shift the burden of carrying stock to the supplier, and Enterprise Resource Planning, which uses software such as SAP in an attempt to track resources within the corporation. Firms using SAP often try to encourage their suppliers to do likewise in an attempt to integrate their entire supply chain. One might also consider the management of consumers by means of loyalty cards and store cards to integrate consumers more closely into the organization (they are the suppliers of demand). The most recent fads in the popular management literature all focus on the increasing cybernetic integration of workers, computer systems and machines. Business Process Re-engineering, Supply Chain Management and Enterprise Resource Planning are all concerned with automating clerical and information processing functions, but they are also aimed at changing the whole structure of the organization. This does not simply require that the organizational processes become automated, but that they become 'obliterated', at least according to the words of management guru Michael Hammer (1990). The most extreme visions of this kind of cybernetic utopia can be found in the literature on knowledge management. Take for example the work of Nonaka and Takeuchi, which states that communication technology can encourage workers or 'crew members' to 'begin constructing a common language and synchronizing their mental and physical rhythms' (Nonaka and Takeuchi 1995: 231). Again, the emphasis of this work is on the primary importance of cooperation and language. The benefits of cybernetic integration are being proclaimed in all areas of management thinking; for example, a recent conference on Customer Relationship Marketing proposed to '. . . bring your people, processes and technology together to build effective customer interaction systems and automate your enterprise' (Porter 2000: 54).

Communication is not simply a means to exploit existing resources, but is in itself of massive value. Media conglomerates are amongst the largest companies on the planet. The hunger for new types of telecommunications is such that the mobile phone company Nokia has grown to dominate the economy of its home country of Finland in a matter of a few years. The importance of communication to imperial control cannot be overemphasized: 'Communication is the form of capitalist production in which capital has succeeded in submitting society entirely and globally to its regime, suppressing all alternative paths' (Hardt and Negri 2000: 347). Deleuze and Guattari (1988) observed that machinic enslavement is now being accomplished through the media and through cybernetic networks of feedback and control. The paradigmatic example they give is power of TV, which sells audiences to advertising firms and other corporations.

Whereas the disciplinary practices of Panoptic institutions were intended to mould the behaviours of individuals, networks of control act through the modulation of their interactions. Whereas discipline operates by segregating and fixing, modulation operates by integrating and organizing differences. Hardt and Negri draw attention to the marketing profession as an exemplary post-modern practice because it embraces difference. Deleuze's masterly essay on the control society observed that the marketing profession is the master of the subtle techniques of control (Deleuze 1995: 181). The slogan of this profession could well be, 'Every difference is an opportunity' (Hardt and Negri 2000: 152). The control society is characterized by increasing levels of cybernetic integration, where all differences are seen as a source of potential exploitation to be integrated into a hierarchy of command and control. The paradigmatic form of modulation according to Deleuze (1995) was to be found in the mechanisms established for the control of money. The move away from a society based on disciplinary power relations to a society based on modulated networks of control is exemplified by the move from the gold standard to floating exchange rates. The control society is engineered according to cybernetic principles of organization such as negative feedback, homeostasis and emergence (Parisi and Terranova 2000). Homeostasis is maintained through negative feedback loops that moderate any potential over reactions to fluctuations in the external environment. Some research has also highlighted that feedback need not be purely reactive in nature and that techniques are currently being refined for anticipating future fluctuations in the environment, by such means as computer simulation and consumer profiling (Elmer 1999; Parisi and Terranova 2000). This could be thought of as a kind of feedback in advance of the event. Automatic systems have been designed specifically to anticipate problems of positive feedback in the world's stock markets and are used as a way of modulating investment decisions. After the rapid stock market crash of 1987, the Stock Exchange Commission in the US implemented 'circuit breakers' that would kick in automatically under exceptional circumstances, notably on the occurrence of a sharp swing in the market and the sudden downward spiralling of share prices. The circuit breaker is designed to suspend trading for a given period of time to allow the equity markets to reassess their information and prevent sustained panic selling. The thresholds that have been set by the Commission before trading may be suspended have been periodically evaluated to determine and improve upon their effectiveness. These steps recognize the instability of unregulated markets and provide a degree of cybernetic re-engineering to help maintain homeostasis within specified limits.

It is through these new communication networks that older forms of sovereign power, such as the nation-state are being undermined, where these same networks also provide the site of production and circulation of capital in the world market. A new form of power is emerging in the wake of the technological advances of the past hundred years. These emerging power relations can be analysed in the kinds of social divisions that are being created and perpetuated by these technologies, most notably in terms of access to education, information and wealth. However, alongside these social divisions there exist forces for increased integration. Work and leisure exist within an increasingly integrated cybernetic system, which brings with it the obligation to connect, to communicate and to consume. Deleuze (1995: 178) remarked on the coming of this new society in the following terms: 'Control societies are taking over from disciplinary societies. "Control" is the name proposed by Burroughs to characterize the new monster, and Foucault sees it fast approaching'. Disciplinary institutions still exist within societies of control, but these are being either supplanted or supplemented by more innovative techniques of control. The control society provides an apparatus of power suitable for the management of the productive forces of immaterial labour. The following section will explore the creative power of the body being developed within this cybernetic system before moving to a summary discussion of power in the network society.

## Cybernetics and superman

> Humanity cannot be improved.
>
> (Virilio 1999: 88)

Back in the eighteenth century, the novelist and satirist Samuel Butler speculated that one day machines would rise to such prominence within society that man himself would become no more than 'a machine tickling aphid'. He proposed that the relationship between man and machine was of the same kind as that which exists between clover and the bees that help to pollinate them. Mankind literally serves as the reproductive organs of the machines. The futurists Alvin and Barbara Toffler have speculated on the potential of humanity, where they predict that soldiers will become genetically engineered and that 'para-humans' will be created to fight our wars for us (Toffler and Toffler 1993: 143). Hollywood has been quick to capitalize on such distopian ideas, and films such as the *Terminator* series or the more laughable *Universal Soldier* have no trouble filling cinemas seats. Whatever the case, it is certainly true that military research is often a testing ground for advances in civilian technology and medicine. Virilio (1986) has described how, during the Great War, the German army was pioneering in its development of prosthetics, and made physical handicaps functional by adapting each man according to his specific disability. As a consequence, it was recommended that deaf soldiers could serve in the artillery and hunchbacks in the automobile corps. A few years after the War, Henry Ford applied a similar kind of reasoning in his new methods of scientific management. In his description of the factory he found that 'out of 7,882 kinds of jobs, 4,034 . . . did not require full physical capacity' (Ford 1924: 108). Ford argued for employing crippled labourers because to do otherwise would be economically wasteful. At that time, it may well have been preferable to work on the assembly line than to be thrown onto the industrial waste heap as a piece of broken and unproductive machinery. Ford even experimented with men bedridden in hospital, claiming they actually recovered quicker if given production line work to do in bed. Here is a revealing explanation of his revolutionary approach to work organization:

> At the time of last analysis of employed, there were 9,563 substandard men. Of these, 123 had crippled or amputated arms, forearms, or hands. One had both hands off. There were 4 totally blind men, 207 blind in one eye, 253 with one eye nearly blind, 37 deaf and dumb, 60 epileptics, 4 with both legs or feet missing, 234 with one foot or leg missing. The others had minor impediments.
>
> (Ford 1924: 110)

In many respects Ford was very enlightened in his inclusive employment policy; however, there was one organ that Ford absolutely insisted should never be used at work, 'above all he [the average worker] wants a job in which he does not have to think' (Ford 1924: 103). There may have been prohibitions against using other organs too, but these are not specified explicitly. He does mention that it was against company policy to employ married women whose husbands already worked for the company, but gives no reason for this. Ford concludes by admitting that repetitive work may be monotonous, but adds that, 'The most thorough research has not brought out a single case of a man's mind being twisted or deadened by the work' (Ford 1924: 106). Unsurprisingly, he never took the opportunity to give it a go and find out for himself.

The unable body and the idea of disability are key to understanding the ways in which the body that has been connected to the new communication technologies (Virilio 1995). The implant of prosthetics has come to dominate the field of surgery in modern medicine. Today, prosthetics are not simply limited to limb replacements, but include the implant of miniature devices, most notably, the use of pacemakers to regulate the heart. The possibility of synthetic organ replacements may be only a few years away. Communication prosthetics for seeing and hearing (i.e. contact lenses and hearing aids) are benefiting greatly from miniaturization and the information technology revolution. The Vice President of Toyota Motor Corporation's research laboratories has boasted, 'Soon we will make microrobots that will take off on a mission to explore the human organism' (quoted in Virilio 1997b: 49). According to Virilio, the body has been transformed into the last territory to be colonized. The extensive transportation and communications infrastructure represented by road, rail, airports, seaports, and telecommunications, will be mirrored by an intensive bodily infrastructure. We have already seen that information warfare is not just a cerebral or virtual affair, and that the spread of certain ideas cannot be disassociated from extant material conditions. A very important element of these conditions is the environment of the human body itself. Marshall McLuhan observed this phenomenon at the outset of the information revolution, explaining that, 'Electromagnetic technology requires utter human docility and quiescence of mediation such as benefits an organism that now wears its brain outside its skull and its nerves outside its hide. Man must serve his electric technology...' (McLuhan 1964: 68).

Our new information technologies require and encourage a state of motility in the body, but where physical movement is reduced to a minimum. Ideal forms of motility include actions such as the channel hopping of the 'coach potato' or 'telly addict' and the double click of the mouse when surfing the web. These technologies, especially the visual ones such as the television and the World Wide Web, have brought into being legions of motile bodies. Virilio has argued that the able-bodied person, who is now equipped to the eyeballs with interactive prosthetics, is modelled on the disabled person who uses similar devices to control their environment. Virilio's work outlines a history of bodily mobility, highlighting how cities have slowed and regulated movement within their boundaries. This slowing down of the population was a crucial factor in the development of capitalism, whereby relatively fixed populations were required to live within cities to man the mills and factories. Today, a new city is being constructed, a virtual 'omnopolis' that links together a mass market of consumers through the television and the Internet. Virilio has equated this immobility with the state of siege experienced by the city dwellers of ancient times (Virilio 1986: 11). Despite this, there are moments when the population picks up speed and becomes revolutionary; for instance, when it took to the streets of Paris in 1789, and on many other occasions since in cities across the globe. We see it today in the protests of Seattle and Genoa, the uprising of the Zapatistas in Mexico and the massive anti-war protests in London. This revolutionary population has been given the name 'the multitude' by Hardt and Negri (2000).

Potentially, the multitude contains everyone, since everyone is subject to exploitation by capital. Even the unemployed are part of the multitude and they are just as important in creating the conditions for the exploitation of a flexible mobile workforce, as are those who have jobs. The poor form the core of this multiplicity since the poor are the elementary productive force of capitalism. The multitude cannot be reduced to a political class; they are beyond representation, where their revolutionary potential lies not in their ability to represent themselves as 'a people', but in their productive forces and creative capacities. Drawing analogy with St. Francis, the authors accord the multitude god-like powers in its potential for creativity, '. . . only the poor has the ability to renew being "The poor is god on earth"' (Hardt and Negri 2000: 157). The fact that the multitude continuously creates the network of communications, which are at once the means of production and the product, is of profound revolutionary import. The networks of the control society are always open to the possibility of resistance, because the effectiveness of these networks is, to a large extent, determined by the freedom of movement allowed within them. This quirk in the design of the networks of control allows for new possibilities of radical democracy, 'the circuits of productive cooperation have made labor-power as a whole capable of constituting itself in government' (Hardt and Negri 2000: 350).

Hardt and Negri's concept of the multitude is not unlike Nietzsche's idea of the 'superman' or 'self overcoming man' (Deleuze 1983). The multitude's continual experiments in living and producing bring about a transvaluation of values. The multitude is the driving force of capital, because the forms of resistance that are spawned by them determine the paths along which capital will move to exploit labour in the future. Hardt and Negri point out that the worker and student protests of the 1960s and 1970s led to a rise in the value of services and intellectual labour, where the movement against factory discipline led to a re-evaluation of the production of culture in terms posed by the counterculture. The re-evaluation of values is not the triumph of heroic individualism but the genius of collective practice. Following the work of Foucault and Deleuze, yet another relationship can be drawn between man and the concept of the superman. The notion that Man is the centre of knowledge and nature had appeared only recently in history and this too will pass. With the Enlightenment, Man took the place of God at the centre of knowledge and nature, but Foucault believed man too would soon be erased by history. This erasure has already begun with the historical forces that have smashed against humanity over the course of the twentieth century using the tools and the weapons systems produced by its factory machinery and its computing machinery. Again, the superman is defined not as a caricature of triumphant individualism, but as a creator of new values. A way of life adapted to prosthetics, cybernetics and genetics, perhaps inconceivable now, will emerge for which new concepts will have to be invented. To quote from the final paragraph of *Foucault*,

> The forces within man enter into a relation with the forces of the outside, those of silicon which supersedes carbon, or genetic components which supersede the organism, or agrammaticalities which supersede the signifier ... As Foucault would say, the superman is much less than the disappearance of living men, and much more than a change of a concept: it is the advent of a new form that is neither God nor man and which, it is hoped, will not prove worse than its two previous forms.
>
> (Deleuze 1988: 131–132)

## The network diagram of power: hierarchies and rhizomes

The diagram of power serves to integrate a broad spectrum of technologies of organization under a specific apparatus of power. In the eighteenth and nineteenth centuries, it was the Panopticon that served to integrate diverse techniques such as confinement, surveillance, dressage, the timetable, and the examination into a single apparatus for discipline. These techniques of organization were disseminated throughout society, where social problems of all kinds – from education, to factory work, to the prison – were all framed in terms of discipline. The argument of this chapter has been that since the dawn of the twentieth century we have witnessed a proliferation of techniques of social control that are not strictly disciplinary, and that we have again crossed a technological threshold where we are now living under a network diagram of power. The network diagram differs in significant respects from the Panoptic diagram, although the Panopticon still thrives under this new apparatus. For one thing, the network functions by facilitating lateral communications between nodes rather than by cutting off such communication. Communication is increasingly seen as a social obligation. It's good to talk. Control is not restricted to fixed disciplinary sites, and operates on a more abstract level, moving within the electronic communication networks that surround us. The network diagram of power has both discursive and non-discursive elements. The non-discursive elements include the visual surveillance of people, their increasing cybernetic integration and an expanding apparatus for the electronic surveillance of data. The discursive elements include those discourses concerning concepts such as the 'flexible worker', 'business process reengineering', 'life long learning', 'the network organization', 'knowledge management' and the globalization of communication networks. A summary of the main differences between the disciplinary diagram of power and the network diagram of power is given in Table 6.1.

The assembly line has been surpassed by the network as the most powerful mode of production (Hardt and Negri 2000). The network is the quintessential post-modern institution, being a virtual space for both the production and the circulation of information. The paradigmatic technological expression of the network appeared in 1964 with the publication of

*Table 6.1* Contrasting the disciplinary and network diagram of power

| *Diagram* | *Disciplinary* | *Network* |
|---|---|---|
| Economic unit | factory | network |
| Labour | material | immaterial |
| Surveillance | visual/panoptic | electronic/panspectric |
| Variation control | moulding body | modulating flow |
| Space | confined cell | connected node |
| Body | docile | Motile |
| Time[1] | timetable | real-time |

Note

1 The concept of real-time has not been examined in any detail in this chapter, but it is such an important feature of the network diagram of power that much of the following chapter is devoted to its analysis.

Paul Baran's work on the design on distributed networks (Baran 1964). In this paper, Baran pinpointed the major weakness of centralized communications systems, notably that all dependent nodes could lose the ability to talk to each other if the central node was subject to attack. Decentralized networks consisting of many interconnected smaller centralized systems have greater robustness, but still rely on the security of a small number of key nodes for their survival. The centralized and decentralized models are in many ways comparable with the Panopticon described earlier in this chapter, where all communications must pass through at least one obligatory point of passage. Baran concluded his analysis by stating that the most flexible kind of network is a distributed network where all the nodes can communicate directly with one another. If communication is the prime goal of a network, then the more redundant channels of communication that are built into the network the better, because redundancy is also a measure of connectivity. The distributed network minimizes the number of obligatory points of passage, and thus makes the network as a whole far more resistant to attack. It is therefore easy to see the importance of the distributed network for the practice of information warfare.

In fact, communications networks are both a social and a technological phenomenon. The idea of a distributed network is not restricted to information technology such as the Internet, and it should be seen as the entirety of connections that make up the world market as a whole. In fact, the world market and its flows of finance, information and commodities, is held up by Hardt and Negri as the ideal type of network. No one exists outside of the network; as Hardt and Negri have expressed this situation, 'In its ideal form there is no outside to the world market: the entire globe is its domain' (Hardt and Negri 2000: 190). The network acts as a series of distributed nodes, through which money, information, people and commodities can pass. Access to these nodes is restricted, yet capitalism itself operates between these nodes, by means of integration and connection,

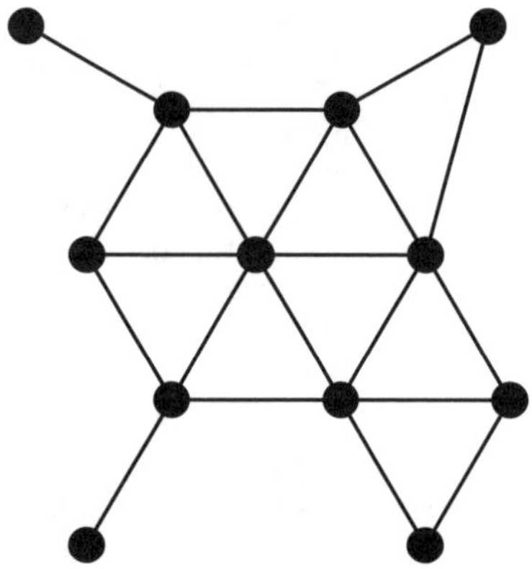

*Figure 6.3* The network diagram (after Baran 1964).

rather than by confinement and separation. Hardt and Negri (2000) have suggested that the new diagram of power has no architectural features, being an entirely virtual space. This may be true to some extent, but it is also the case that restricted access within these networks represents a new hierarchy. Zuboff's early analysis of the effects of information technologies found that, 'As the electronic text becomes the symbolic surrogate for the organization's vital activities, access rules become the fundamental equivalent of organization structures' (Zuboff 1988: 356). Hardt and Negri's own description of the social divisions of capitalism outlines a similar phenomenon that is written into the fabric of society itself, in terms of social hierarchies and access to knowledge and wealth. A topological representation of the diagram of forces that underpin the network society is described in Figure 6.3.

In the terms given to us by Deleuze and Guattari (1988), the network is neither a tree nor a rhizome, but both trees and rhizomes can grow within it. We cannot predict what new products or forms of social life will emerge from the development of our communication networks; we form a rhizome with them as we engage in collective experiments, as embodied in communities as diverse as Napster, Linux, and the Zapatistas to the more general conceptualization of the 'multitude'. Yet, in the wake of these rhizomic experiments, trees are forming, with the restricted access rights to information, intellectual property and wealth.

# 7 Real-time

## Control and resistance

> The element of 'time' is the heart of the fundamental difficulty of nearly all economic problems.
>
> (John Maynard Keynes)

> This is the first thing I have understood
> Time is the echo of an axe in a wood
>
> (Philip Larkin)

This chapter shows how different conceptions of time are bound up with different forms of information technology and methods of social control. Resistance to such control is often characterized by attempts to evade institutional methods of time keeping, such as the timetable and real-time control, and to establish a sense of identity independently from these mechanisms. The argument begins with a critique of a strictly quantitative conception of time, following Marx's analysis of money. This analysis goes far beyond platitudes like 'time is money', and examines the social relations and techniques of control surrounding the equivalence between time and money. The work of Antonio Negri is then used, which supplements and reworks Marx's analysis in line with the productive technologies of immaterial labour. Negri's work shows that with the rise of the service sector, the entire social fabric has become subject to exploitation, whether in terms of the production of physical commodities or the generation of ideas and emotions. This he terms the 'real subsumption of society under capital', where the entire life of labour becomes subject to measurement, control and exploitation. The most recent symptom of the subsumption of society is embodied in technologies of real-time control, which constitute an apparatus of command that has emerged simultaneously in both civilian life and the military field of operations. The final sections of this chapter are devoted to a re-evaluation of real time as a qualitative concept, which is in many ways critical of the purely quantitative concept of time. These sections draw on the critique of time outlined by Nietzsche, Bergson and Negri, which view time as a fundamentally creative concept rather than merely an instrument for measurement and control.

The writer William Burroughs has observed that time has always been an essential element in any system of control; in Burroughs' words:

> A basic impasse of all control machines is this: Control needs time in which to exercise control.
>
> (Burroughs 1986: 117)

Time and control are tightly bound together. From the earliest days of the factory system, the clock has served to coordinate and regulate work. Prior to the factory, monasteries used timetables to order the day's activities around various times for prayer spread throughout the day. Burroughs' own literary experiments with the concept of time drew upon the ancient Mayan society, where priests possessed restricted knowledge of the solar, lunar and ceremonial calendars through which they could organize their agriculture and thereby control their people. The twentieth century has seen the development of many temporal techniques for control including, the time–motion study, Taylorist work control, Just-In-Time (JIT) production, and Enterprise Resource Management software. The relatively recent concept of 'real-time' control is the latest in a series of sophisticated techniques for organizational control, and has developed with the rise of the computer within society. These examples all combine a particular concept of measurable time with a particular regime of control.

## Time and control 1: money

Thinkers as diverse as Nietzsche and Marx agree on the importance of money and its distinctive function as a symbol. According to Marx, in the commodity form money serves as the ultimate symbol of human activity: '. . . every commodity is a symbol, since as value, it is only the material shell of the human labour expended on it' (Marx 1976: 185). Nietzsche saw money as a kind of primordial symbol, and he posed the problem in the following way: 'Fixing prices, setting values, working out equivalents, exchanging – this preoccupied man's first thoughts to such a degree that in a certain sense it *constitutes* thought . . .' (Nietzsche 1994: 49, emphasis in original). Nietzsche's genealogical studies highlight the fact that both thought and money establish a process of social exchange, and historically they appeared to have developed side by side.

For Marx, time was the defining feature of labour, it is the medium through which labour-power becomes quantifiable: 'The worker is nothing more than personified labour-time' (Marx 1976: 352–353). But something happens to the concept of time when it becomes a medium through which labour-power is measured, that is it becomes homogenized and, as Marx repeatedly states, 'congealed'. The nature of the exchange relation allows this to occur, as different products of a day's work are compared and exchanged. In Marx's (1976: 186) words, '[exchange] value is determined by the labour-time required for its production, and is expressed in the quantity of any other commodity in which the same amount of labour-time is congealed'. This idea of congealed time suggests a process of solidification and homogenization, by which labour-time can be quantified and measured. The clock, the setting of working hours and the division of labour are all technologies that Marx lists which facilitate this process of congealing. This understanding of time has infiltrated our consciousness to such an extent, that we can now say without danger of sounding ridiculous that 'time is money'.

This mechanism of abstraction moves to a new level as money is used as the common medium for exchange. The relationship between money and time is established through the process of measurement and exchange, 'Money as a measure of value is the necessary form of appearance of the measure of value which is immanent in commodities, namely labour-time' (Marx 1976: 188). When labour-time is expressed purely in terms of money, Marx uses the term 'crystallized' rather than 'congealed', suggesting another dimension in the homogenization of time (Marx 1976: 181, 183, 187, 229, 297). Money is the universal form of equivalence and the perfect medium of exchange. As such it possesses certain fundamental properties, 'Only a material whose every sample possesses the same uniform quality can be an adequate form of appearance of value, that is a material embodiment of abstract and therefore equal human labour' (Marx 1976: 184). In order to exchange objects of different magnitudes of value 'it must therefore be divisible at will, and it must also be possible to assemble it again from its component parts' (Marx 1976: 184). The money form has certain crystalline qualities since each of its parts is both uniform and divisible, and when considered as a whole it is composed of identical samples (Calasso 1995). This process of metamorphosis between money and the commodity is analogous to crystallization since it transforms all commodities into one durable and uniform crystal. The money crystal can then be dissolved into a means for purchasing products and services. Although money is the semiotic basis for exchange, it is time not money that renders commodities commensurable and exchangeable. Money represents labour time when this time is considered purely in terms of its socially average character. This happens when labour is itself exchanged for money at a given point in time and place, depending upon the relative supply and demand for that labour. There are also some things that can afford a price which have no labour-time embodied in them, such as honesty, conscience or honour. These are sold without being worked for.

Much of Marx's analysis concerns the various ways in which the capitalist attempts to exploit more and more of the labourer's time, by means of lengthening the working day, haggling over meal times, extending the working week, and punishing tardiness with fines which often exceeded wages. The careful manipulation of time was essential to the development of capitalism. When child-labour laws were introduced in the nineteenth century, limiting the working week, a relay system was introduced by many manufacturers whereby working children were moved between different factories rather than returned to rest in their homes. The invention of the machine, paradoxically, led to an increase in pressure rather than a lightening of the workers' load. Increased efficiencies meant a greater proportion of a day's work went to the capitalist as surplus labour, and the cost of leaving a machine idle was such that it required continuous superintendence day and night. Marx defined the machine as 'a means for producing surplus value' (Marx 1976: 492) in contradiction to the myth that technology reduces the working day (Marx 1976: 533). On the one hand, the machine of the industrial revolution demanded that human habits be adapted to its own monotonous rhythm, and on the other hand, attention to clock time was required to synchronize the efforts of labour in large-scale manufacturing. The historian E.P. Thompson has observed with characteristic irony that, 'for fifty years of disciplined servitude to work, the enlightened employer gave his employee an engraved gold watch' (Thompson 1967: 70). Marx goes into minute detail showing exactly how capital exercised vampire-like powers over the workforce of his time, sucking the life blood out of them, leaving nothing behind but a hollow husk. He observed that despite capital's overwhelming concern with the regulation of time, there was one vital question that it never asked: 'Capital asks no questions about the length of life of labour-power' (Marx 1976: 376). Marx's analysis of capital, therefore, has two very important implications for our concept of time. First, he showed that control needs time, and described in detail the many and varied strategies thought up by capital for this purpose. Second, he showed that our understanding of time is itself transformed, congealed and crystallized, in connection with other concepts, especially those of labour and money.

Marx was probably the first thinker to recognize the importance of time for the development of techniques of social control. There are many points in Marx's analysis of capital that directly mirror Foucault's analysis of disciplinary power, where each analysis is concerned with technologies of time. In Foucault's study of disciplinary power, the first technique – the discussion of the control of activity – begins with an explanation of the crucial role played by the timetable, which is commonly used throughout the disciplinary apparatus of society. To waste time is immoral and requires a corrective provided by the disciplinary apparatus. The timetable is useful for coordinating activity, but also for allocating and measuring the duration of each activity throughout the day. The second technique for the control of activity is concerned directly with time in the temporal elaboration of the act (Foucault 1977). This involves breaking every activity down into its elements, calculating the time needed to perform each one, and controlling each in succession. Time is therefore essential for the exercise of control where, 'Time penetrates the body and with it all the meticulous controls of power' (Foucault 1977: 152).

## Time and control 2: the real subsumption of society

> In the period of the man–machine command becomes biopolitical control. What is meant by 'biopolitical control'? It is measure (that is, organization and limit) of the time of life.
>
> (Negri 2003: 257)

Negri (2003) has developed Marx's analysis of time, updating it for a new mode of production, notably that of immaterial labour. His analysis agrees with Marx's, that time does not only measure labour, but reduces it to a homogeneous substance. By making labour more or less homogeneous, different forms of labour could be compared and exchanged, in terms of their relative productivity, where such productivity was in turn measured in terms of time. Negri has distinguished between two practices of time: analytic and productive. The analytic practice of time is defined as those processes of equivalence and exchange that allow capital to circulate. Negri also called this the time of command, which is effected by means of measurement. Time becomes a quantitative measurement, an instrument that exists solely for the purposes of social control. In Negri's own words, the analytic practice of capital is 'the parasitical organization that seeks to bestow measure on the immeasurable' (Negri 2003: 260). This process of equivalence and measure is not given as a natural consequence of human cooperation, but is the effect of coercion. The time of command lacks its own creative capacity, and the essence of this practice of time is clearly reflected in Marx's metaphor of 'dead labour'. In contrast to the analytic of command, the practice of productive time is defined in terms of cooperation. Cooperation is the essential human activity from which all other productivity and creativity spring. Productive time derives from 'living labour', the production of cooperation from which all subsequent forms of organization emerge. Whereas the time of command is encapsulated in the exchange-value of a product, productive time is encapsulated in its use-value, which exists prior to the establishment of an exchange value.

Negri develops a theme originally set out by Marx, which they describe as the real subsumption of labour by capital. In short, this involves the insinuation of capital into the fabric of society. During the nineteenth century the subsumption of society included the struggles over the length of the working day, the exploitation of women and children as well as men, and the apparent disregard for the health and lifespan of the labourer. Marx also observed that work in factories and on machines could seriously stunt other potential abilities, especially one's intellectual powers. This exposed a limit to the penetration of capital into social relations at that particular time in history. In the period of early capitalism, the labourer's intellectual and emotional capabilities seriously suffered as a result of the repetitious and monotonous work; the labourer's capabilities were stunted but were not directly exploited by capital. With the rise of immaterial labour and forms of labour where emotional and intellectual products constitute the key productive apparatus of society, the real subsumption of society under capital is complete. Following in the footsteps of Foucault, Negri has shown that under such conditions our very subjectivity is at stake, constituting as it does the productive power of society. In Negri's own words, 'The conditions for Marx's notion of exploitation are over, because industrial production is no longer fundamental and has become only a simple consequence of the productive activity of the biopolitical base' (Negri 2003: 233).

The invention of 'real-time' is yet another technological symptom of the subsumption of society under capital. Real-time control has been used to make labour more time-efficient in many different ways, for example, by 're-engineering business processes' to cut down levels of hierarchy and layers of middle managers, by instituting Just-In-Time (JIT) management to reduce stock levels, and using Enterprise Resources Planning (ERP) software to keep track of all corporate resources and finances in 'real-time'. This issue of real-time control and its relation to information warfare will now be explained in greater detail.

## Information warfare and real-time control

> The loss of material space leads to the government of nothing but time.
>
> (Virilio 1986: 141)

Real time is the time of information warfare. The concept of 'real-time' has emerged from the IT industry and it is essentially a computer-mediated concept of time. It implies the capacity of computers to interact with events in the world as they happen. Technologies for real-time control have revolutionized the organization of civilian and military operations simultaneously. Over the past few decades, business operations have been subject to increasing cybernetic integration using real-time technologies to underpin business activities, all the way from procurement, to the manufacture of the product or service, to sales and marketing and the relationship with potential and actual customers. In the terms of the Harvard economist, Michael Porter (2001: 71), 'By easing and speeding the exchange of real-time information, it [the Internet] enables improvements throughout the entire value chain, across almost every company and industry'. Porter has observed that information technology has always had a profound effect on the value chain of corporations, because every business activity requires the creation, processing and communication of information (Porter 2001: 74). Whereas the earlier developments in computing technology led to massive increases in processing power, the more recent developments have led to greater connectivity and a proliferation of links between the different elements of the value chain, using tools such as Supply Chain Management, Customer Relationship Management and Enterprise Resource Planning. Virtual activities do not eliminate the need for physical activities, but they serve to modulate the interactions between these activities.

The 1991 Gulf War was arguably the first major conflict fought in real-time, with extensive use of satellites to coordinate decisions and actions between the United States, Europe and the Middle East. Ideally, real-time control technologies have almost no interval between an event, its virtual representation and the programmed response to it. Real-time technology has enabled the globalization of the financial and commodity markets on a scale not seen before. Alongside the creation of a globalized time, there has been an expansion of international markets and the dissolution of cultural borders to some extent. One of the most striking manifestations of real-time is the Internet, the major part of which is represented by one language, English. To be wired into this symbol of wealth and modern technological progress means to be able to communicate in English. Real-time has had the effect of dissolving some of the boundaries between regional cultures, along with the particularities of the real spaces of those cultures (Virilio 1997b). With the development of real-time control technology, it may be said that the new borders are defined more in terms of time than of space (Virilio 1999). This has entailed a redrawing of the geography of international business, with manufacturing industry being exported to less-developed countries. It has also involved a change in the time in which work is done, where traditional distinctions between work and home are breaking down with the advent of flexi-time and telecommuting, and the working day has developed into a 24/7 society. The new frontiers of capitalism are embodied in the concept of real-time, which allows for the integration and exploitation of networks of people and machines on a new communicative level (Hardt and Negri 2000). At this level, capitalism is no longer restricted to alienating people from the product of their labour, but extends into communicative networks, into our use of language (e.g. branding and marketing) and into the knowledge and ideas people generate (e.g. intellectual copyright). Capital has always been concerned with the control of the labourer's time, but the invention of real-time has facilitated a further intrusion of capital into the lifetime of the worker.

Real-time technology has also spawned its own specific form of resistance, which has come to be called 'hacking'. Hacking can involve attempts to disrupt real-time through disruptive techniques such as the computer virus, but it also includes creative online communities like Linux. The hacker ethic may be defined as that of the enthusiastic amateur, whose motivations are not primarily monetary, but involve the sharing of information within a community of enthusiasts, the creation of something that they find intrinsically valuable, and the facilitation of access for everyone to information and computer resources (Himanen

2001). Quite simply, the hacker ethic is about the creation of an online commons for information, in clear distinction to the production of private property. Hackers are a kind of double to the professional flexible worker. The flexible working practices that have emerged with the growth of the information economies have been achieved at the cost of greater levels of social exclusion. Huws (2003) has observed that flexibility and social exclusion share a common set of symptoms, including high unemployment, a rise in short term contracts, and the loss of social protection for workers. The amateur hacker is quite the opposite of those professionals who have adopted flexitime to schedule a private life into a heavy working week. The hacker ethic is a rejection of the Protestant work ethic, disrupting the whole idea of a working life defined in terms of labour time. As Hinamen has explained, 'our life is here and now . . . Hackers do not subscribe to the adage "time is money" but rather to the adage "it's my life"' (Hinamen 2001: 40). Hackers are fully aware of the fact that real-time has brought with it a new regime of control, but hackers also embody a basic impasse of this new regime of control, i.e. that time is always needed to exercise control and this gap can itself be disrupted from anywhere within the network.

Real-time control has also been employed by the military as a new weapon against guerrilla tactics. Guerrilla warfare has always exploited the local environment and a variety of disruptive techniques so that guerrilla fighters might move unimpeded in their own time. Nation-states are learning from such tactics in their use of global satellite and communication networks to wage a real-time war in regions like the Gulf, the Balkans, Afghanistan and Mexico. Real-time control allows the military more effective coordination of operations based on special forces and task forces, and also to intensify surveillance of enemy territory, hence removing many of the advantages of traditional methods of guerrilla warfare. The 'netwars' described by Arquilla and Ronfeldt (1997) are themselves a radical innovation in the use of real-time technologies to manage conflict. Guerrilla groups have also been adapting these technologies to their needs, most notably in the case of the Zapatista rebels in Mexico. The relatively unregulated and decentralized power of the Internet provides such groups and other non-state actors with the opportunity to present their own information or misinformation, which can be set up in real-time to counter the propaganda that may be broadcast by the traditional news media. This netwar does not simply involve the exploitation of the Internet, but connects a network of both local protest groups and outside non-governmental organizations to help support the aims of their netwar. Information warfare is thus learning from and changing the nature of guerrilla warfare by creating its own local real-time propaganda in combination with disruptive techniques against the state's communications networks. The revolution in military affairs described by RAND analysts (Arquilla and Ronfeldt 1997) is an attempt to wrestle with the kinds of innovations being adopted by local protest groups and practitioners of information warfare, like the Zapatistas. Characteristic of the above discussion of time is the idea that real time is homogeneous and measurable. In contrast to this, the following sections will explore alternative conceptions of time that build on a qualitative appreciation of time. This argument will follow Nietzsche, Bergson and Negri, all of whose works highlight the crucial role of creativity in their conception of time.

## The immeasurable: Nietzsche, Bergson and Negri

### *Nietzsche: the creative time of history*

Nietzsche was an early critic of straightforward quantitative, linear conceptions of time. His work analyses the concept of time in a number of different guises, as memory, as history and as becoming. Memory and history may be understood as basic information technologies, the former allowing access to personal information about the past and the latter involving more collective access to past events. Nietzsche reconceived time not as an *a priori* category of existence, but always as the result of interpretation.

Nietzsche did not accede any transcendental status to time, but viewed time as a powerful metaphor that has come to predominate the way we think. There is no time per se, 'space and time do not exist in themselves' (Nietzsche 1967: 293). To understand this proposition, a short explanation of Nietzsche's concept of becoming will be useful. The world exists in a state of continual change and variation, where the ontological reality of the world is a process of becoming without any given goal. Can there be any knowledge of this process of becoming? In fact, knowledge does not pertain to becoming, but to entities that are abstracted from this process of becoming, captured by language in the form of concepts (Nietzsche 1967: 298). Knowledge thus presupposes that things exist which are stable over time and are regularly recurring. However, we can only distinguish between different things by ignoring the infinity of differences of which the world is made. This requires a huge creative effort on the part of human kind, in Nietzsche's own words, 'Before there is "thought" there must have been invention; the *construction* of identical cases, of the appearance of sameness, is more primitive than the knowledge of sameness' (Nietzsche 1967: 293, emphasis in original). Knowledge of the world is only possible under the condition that identical cases occur in reality. For identical cases to exist one must first invent the metaphor of the 'thing', which is defined in terms of an unchanging essence that shares a qualitative equality with other identical things. The invented metaphorical status of the thing has been entirely forgotten. This point can be illustrated if one considers how we identify a thing: 'The properties of a thing are effects on other things: if one removes other "things," then a thing has no properties, i.e. there is no thing without other things, i.e. there is no "thing-in-itself"' (Nietzsche 1967: 302). The concept of the thing-in-itself is metaphorical in origin, but even so, it is an essential condition for the creation of knowledge. Nietzsche traced the origin of the metaphor of the 'thing' to the concept of the ego, 'Man ... derived the concept of being only from the concept ego, he posited things as possessing being according to his own image, according to the concept of ego as cause' (Nietzsche 1990a: 60–61). He was perhaps the first philosopher to see that thinking is profoundly anthropomorphic and conditioned by the historical contingencies of language. In this respect, our habit of blaming things for our own mistakes and misfortunes is hardly surprising, since we see little egos everywhere.

Nietzsche accorded the idea of becoming (metamorphosis) primacy over that of being (things). Becoming, therefore, is not simply a change of state in things, because this would already presuppose that a stable unchanging thing existed prior to the flux of becoming. The process of becoming should not be understood as a thing that changes, becoming is change itself. As such, knowledge does not really represent reality, nor does it tell us anything about the process of becoming; it is merely the imposition of logic and mathematics onto our experiences in an attempt to categorize and control them. We have forgotten the metaphorical nature of this process, as Kofman says, 'Metaphorical activity, always already forgotten, is secondarily repressed by being deliberately abandoned in favour of the concept, of logic and science' (Kofman 1983: 43). The quantification and measurement of things requires the imposition of metaphors from the realm of mathematics onto the qualitative realm of experience. The concept of a quantifiable and measurable time is therefore already an interpretation; it is nothing more than the metaphorical imposition of mathematical concepts (numbers, linear succession, counting, etc) onto the physical world of becoming. Nietzsche builds on this to develop a primarily qualitative conception of time by examining the specific processes through which we develop our knowledge of time, that is, through memory and history. Memory and history are not simply abstract concepts, but are complex information technologies through which we experience and understand time.

The animal grazing in the field has no awareness of the passage of time; it lives in a continual state of forgetfulness (Nietzsche 1980). Nietzsche tells us that consciousness of time is essential to consciousness itself. The fundamental building blocks of culture involve techniques for developing an awareness of time, such as memory and history. These techniques provide ways of fixing events in our consciousness. These events do not stay in the memory of their own accord, as if they were simply slotted into a vacant hole in the brain; they need to be 'burnt in'. Pain is a primary force in memory, in the development of our consciousness and our awareness of time. Nietzsche lists numerous techniques that have contributed to the development of memory over the ages. In this list, he includes religious ceremonies, sacrifices (e.g. of the first born), promise making, penal law ('don't forget that you are responsible for your deeds and your debts'), and a host of ascetic practices that are fundamental for producing a body in which ideas can be retained (Nietzsche 1994: 41–42). In his focus on material social practices, Nietzsche's analysis of time has some significant points of similarity to that of Marx, where both reveal how particular material conditions can change our understanding of time. Nietzsche's analysis focuses on how the human animal becomes conscious of time, including the ascetic practices by which memory is forged and through the concept of history. In contrast, Marx was not concerned with memory as such; nevertheless, his concept of labour-time is bound up with the material practices of measurement, exchange and control, which tied the human body to a new rhythm with the invention of the clock and the machine.

The concept of memory also brings with it the practice of forgetting. We must be able to stop remembering what has just happened in order to act and live. The active and creative person possesses an excellent memory and an equally powerful capacity for forgetting: 'all acting requires forgetting' (Nietzsche 1980: 10). The man who cannot forget would feel the overwhelming burden of time, 'he sees everything flow apart in mobile points and loses himself in the stream of becoming' (Nietzsche 1980: 9). Forgetting is therefore an essential element of activity and creativity, whereas the process of remembering tends to emphasize what is already familiar to us. Remembering is a process of classification and pigeonholing, where 'the drive to make equal has already been subdued' (Nietzsche 1967: 274). Our desire for knowledge and capacity for memory develop hand in hand. The capacity for memory increases at the same time as our awareness of change and difference diminishes, 'memory is possible only with a continual emphasising of what is already familiar, experienced' (Nietzsche 1967: 289). Memory can therefore be seen as a relatively reactive force, which demands an equal power of forgetfulness for us to become active and creative. These plastic forces of life are also discussed in some depth in Nietzsche's work on history – our collective memory.

For Nietzsche, there was no definitive past, since history is always conceived in terms of the present: 'The past always speaks as an oracle: only as master builders of the future who know the present will you understand it' (Nietzsche 1980: 38). Nietzsche discussed three ways of doing history, each of which reinforces a passive appreciation of time and prevents us from using history for the purposes of life. Historical knowledge can be developed in one of three ways: monumental history, antiquarian history or critical history. Monumental history celebrates the great moments of the past as events that are worthy of imitation, and yet are fated never to be repeated (Nietzsche 1980: 17). This conception of the past tends to overgeneralize and equate differences. This kind of history becomes particularly pernicious in the hands of the 'impotent and inactive'. One might take, for instance, the way Napoleon looked on the Roman Empire as his historical antecedent, the way Goebbels used a variety of mythologies in justifying the racist policies of the Third Reich, or even the way Bill Clinton modelled himself on John F. Kennedy. Marx also mocked this kind of simple-minded manipulation of history in his account of Louis Bonaparte's revolution of 1848: 'Hegel remarks somewhere that all facts and personages of great importance in world history occur, as it were, twice. He forgot to add: the first time as tragedy, the second as farce' (Marx 1934: 10).

Nietzsche also criticized 'antiquarian' and 'critical' approaches to history, which he saw as uncreative and weakly accepting the burden of the old values that had been handed down to them by their ancestors. The antiquarian historian is lost in veneration of ancient peoples at the cost of a complete rejection of new and developing powers. This kind of history is able to preserve the past, but has no life of its own, no creative powers. History becomes a dusty and academic enterprise, with little relevance to the present. Again, this type of historian lacks any active or creative powers. The third type of history, which presented critical understanding of the past, views it in all its bloodstained detail, and finds it guilty of the most appalling crimes. This approach to history emphasizes the dangers inherent in action, which has a paralysing effect on its students. Students of critical history become afraid to live actively and dare not risk making a history of their own. Nietzsche explained this danger in more detail with reference to mankind's creative capacity,

> If no constructive drive is active behind the historical drive, if one does not destroy and clear away so that a future, already alive in our hope, may build its house on the cleared ground, if justice alone rules, then the creative instinct is enfeebled and discouraged.
>
> (Nietzsche 1980: 38–39)

Nietzsche's essays are not a critique of history per se, but particular approaches to history, which tend to neglect the prejudices and the way of life valued by the historian. The three approaches to history he discussed all have their own evaluations of the present and are all influenced by the particular material practices in which they are grounded. Furthermore, they are thoroughly incompatible with each other, 'Each of the three kinds of history is justified in only one soil and one climate: in every other it grows into a noxious weed' (Nietzsche 1980: 18). Nietzsche's work shows us that to arrive at a straightforward linear conception of history requires a host of serious misunderstandings. For one thing, it requires the existence of stable points in the past, which we can follow by tracing a path backwards through history, and yet we find that the selection of these points is always open to interpretation. It is hardly contentious to point out that history is written by the victors. Every event is, first, an interpretation, it demands of us a perspective on the past that may only be understood in terms of our relation to the present and our attempt to master it. The past is not a linear series of events because it always serves life in some way; it serves either to enhance or diminish its active forces.

Nietzsche was ultimately concerned with the problem of value and how, living on the edge of the abyss, human society has created such value. His genealogical studies can be defined precisely as an investigation of the origin of values and the value of origins (Deleuze 1983). Nietzsche wrestled with the peculiar capacity for creativity that lies within the human animal, proposing that our awareness of time actually cuts us off from our active and creative forces. Man is weighed down by the past and his supposedly timeless values of good and evil: 'Almost from the cradle we are presented with heavy words and values: this dowry calls itself "Good" and "Evil". For its sake we are forgiven for being alive' (Nietzsche 1969: 211). In contrast to this overburdened creature, the creative man is the kind of man who has the power of forgetting, to be released from the burden of the present. For Nietzsche, to affirm life is not to accept responsibility for the present, but to create, and to unleash the labyrinthine forces of the unconscious (Deleuze 1983). The creative type is open to the creative potential of the future; he is the 'self-overcoming man' or superman. Nietzsche thus argued that there is no time per se, but a variety of conceptions of time, which are inextricably bound up with the creative forces of the present. Time is not merely a quantitative measure and an instrument of control, but is a fundamentally creative force that is necessary for any self-understanding at all, both in terms of the way we constitute our memories and our collective history.

### *Bergson: the creative time of duration*

Henri Bergson shared many of the same issues that Nietzsche was concerned with: they were each proponents of a philosophy of becoming, and they each developed non-spatial, qualitative conceptions of time. Bergson attempted to develop a novel concept of time that he called 'duration', which was not defined spatially in terms of extension or measurement but represents a dimension all of its own. Why was Bergson so concerned with creating a new concept of time? Following Zeno, he argued that a spatialized, quantifiable concept of time presents us with an untenable paradox. Movement in space is only possible as time passes, but if time itself is really spatial (i.e. quantifiable and measurable), then no other dimension exists that can allow it to pass. Consider the flight of an arrow through a number of points in space. At a given instant, the arrow occupies only one position in space, but for the arrow to occupy that space at that instant it must be at rest. The arrow, being a certain length, cannot stretch itself forward into the next section of space in the following slot in time, and so it can never move. Another similar paradox that arises from the spatial view of movement can be defined when the movement of an arrow proceeds towards its target, passing though points in space forever halving its distance to that target. However, no matter how close the arrow gets, it will never reach its target. Movement, therefore, must have a non-spatial dimension to it and this dimension is what Bergson called 'duration'. Duration allows time to pass; in Bergson's own words, it is 'the continuation of what no longer exists into what does exist' (Bergson 1999: 33).

Duration is a concept of time, which is not a transcendent category or an external measure, but is immanent to things. Deleuze has defined it as 'a becoming that endures, a change that is substance itself' (Deleuze 1991: 37). It would be a mistake to see duration as a purely social phenomenon because it concerns the very nature of physical reality. The examples of movement that Zeno and Bergson employ are not social systems but physical ones. The problem Bergson identified is that mathematics does not provide us with an adequate concept of movement or duration. Adamson (2000) has described this problem in terms of the nature of continuity. In mathematics continuity is defined in terms of a series of points that are themselves discrete. Continuity itself is to be found in the infinitely small interval between these points and it is precisely this interval that cannot be divided. The same can be said of Bergson's concept of duration since it is not a spatial and quantitative dimension and must be understood as a purely qualitative concept. Whereas space is defined by an underlying homogeneity in things that allows for quantitative differences of degree, duration is defined by an underlying qualitative variation that allows for differences in kind. Duration can be divided but only when it produces a difference in kind; it is fundamentally creative. Like Nietzsche, Bergson found that time cannot be reduced to a quantitative measure, but is a generative force in its own right.

## *Duration, cinema and invention*

> Knowing how to look is a means of inventing.
>
> (Salvador Dali, *Oui*)

Bergson criticized cinema on the grounds that it provided the perfect illustration of false movement, because it presents us with the illusion of movement constructed out of many immobile sections. The speed at which one frame replaces another on screen is sufficiently fast that it is not consciously recognized by the human eye, thus giving the impression of continuous movement on screen. Cinema is therefore built on the same kind of paradoxes that Zeno outlined some millennia previously. Despite Bergson's rejection of cinema, Deleuze (1992) has argued that the evolution of cinema has in fact overcome its initial limitations and is a creative art form that has the capability of providing us with important lessons in how to look at the world.

Cinema was born at the dawn of the information age, and in many respects its methods of organization are representative of the methods of organization of this new age. The potential of cinema to play a formative role in people's perceptions was recognized from the outset by its powerful sponsors, and it has always provided a key instrument for propaganda. The propaganda element of a film may be quite explicit, such as Leni Riefenstahl's hymn to Nazi ideology, *Triumph of the Will*, or it may enter a movie on a more subtle level, such as in Ridley Scott's *Black Hawk Down*, which portrays the misguided incursion of US forces into Somalia in 1993 where 18 of its servicemen were killed. Although this latter film does not explicitly extol any systematic ideology, it does serve as an apologia for the invasion of a foreign country, and whereas we are given a portrayal of the humanity of the US servicemen who lost their lives, no such insight is given into the many more Somali's who were killed by the invading force. Cinema is a machine for the propagation of propaganda and social stereotypes of all kinds. Occasionally cinema can also play a critical role in undermining such stereotypes through parody and burlesque, most notably in Chaplin's film such as *The Great Dictator*, *Modern Times* and *Monsieur Verdoux*. Chaplin's films reflect upon the social conditions in which we live and show us just how intolerable some of these situations really are. Chaplin's films do not have to lecture us on the evils of modern times and his early films illuminated the absurdity and hypocrisy that his times had created by images alone. In his later films, he did use the latest technology of the talkies and Deleuze has explained the force of these films in the following terms:

> What the discourses say in *The Great Dictator* and *Monsieur Verdoux*, is that Society puts itself in the situation of making any powerful man into a bloody dictator, any businessman a murderer, literally a murderer, because it gives us too much incentive to be evil, instead of giving rise to situations where freedom and humanity would be bound up with our interest or our *raison d'etre*.
>
> (Deleuze 1992: 172)

Deleuze observed that cinema provides a way of seeing that marks a clear break with previous technologies, such as photography or painting. The photograph is a snapshot in time that might be described as a 'temporal mould'. In contrast, the cinema is not restricted to the snapshot of time, but serves instead as a means of modulating between shots, allowing for a far more mobile technology of perception. This difference between techniques for moulding and modulating is also a crucial distinction in the kinds of political technologies of disciplinary societies, based on moulding, and the political technologies of the post-disciplinary control society, based on modulation. Given the crucial importance of modulation to the functioning of the control society, the example of cinema may prove to be a fruitful basis for analysing the new techniques of control and creative forms of resistance to such control.

At the conclusion of Deleuze's first book on cinema he observed that images tend to make use of clichés. Even the great masters, such as Hitchcock and Chaplin, made use of clichés, the former undermining them through horror and the latter through parody and burlesque. Deleuze leaves us with the question, is it possible to create an image that would not resort to cliché? Can the cinema form a critical role in the way we think and provide a way of looking at the world which is at once aesthetic and ethical? This question will now be addressed by drawing on the work of the director David Lynch, specifically his minor masterpiece *Lost Highway*.

The film starts with our protagonist, Fred, taking his first drag on a cigarette, setting himself up for the day. He hears the door buzzer go, walks over and presses the talk button. A voice at the door says, 'Dick Laurent is dead'. The name doesn't seem to mean anything to him. When he checks the window, he can see no one at the door, nor is there anyone on the street. He is puzzled. Fred and his wife then receive a series of packages that contain video footage that appears to have been shot from inside their own house. One of these unwelcome films ends with a shot of them asleep in their bed. The motives behind these packages are far from clear, especially given that they were sent to the presumed victims. They call the police, who are helpless to do anything given the lack of any obvious leads. Shortly after this incident Fred receives another package, it seems the same as the others except that this time the video finishes with a shot of Fred standing over the mutilated body of his wife. Fred is arrested for her murder and is subsequently imprisoned. Video footage is discovered which seems to show Fred killing his wife, but despite this, he maintains his innocence. After having spent some time in prison awaiting execution, he starts complaining of headaches. A guard checks on his cell one day and discovers that he has disappeared and that another much younger man has taken his place, Peter. The authorities are stumped and are forced to release their new inmate who goes home to his parents unable to explain how he suddenly appeared in a prison cell. Later in the movie, the younger man falls in love with the fiancée of a man called Dick Laurent. Dick Laurent is a powerful menacing force in the movie. He is a gangster of some kind and seems to have known the murdered woman. His fiancée also appears to be the double of the murdered woman, perhaps her twin. Peter and the woman begin an affair, hidden from Laurent. The young man discovers that Dick Laurent makes pornographic films, in which Laurent has forced his beloved to participate. The lovers plan to escape the reach of the evil Laurent. Shortly afterwards he undergoes another transformation back into the condemned man, Fred. It was Laurent who had killed Fred's wife, even though the video evidence suggests otherwise. Fred tracks down the evil Laurent and kills him. He then returns to his own house, hits the door buzzer and says, 'Dick Laurent is dead'. In the final scene he is fleeing from the police, speeding down the highway and he begins yet another transformation.

All the possible transformations within the storyline of this movie are too numerous to list. It could be argued that for all films the attempt to portray a linear narrative or an impression of continuity is already a deception, founded on stills, cutting shots and an infinite number of possible viewpoints. Despite this, we are in the habit of suspending our disbelief and perceiving a semblance of linearity within a story so long as the temporal sequence of events unfolds in order, even despite the frequent temporal and spatial jumps that a film may require us to make. In the case of *Lost Highway*, however, the temporal sequence of events is undone and our viewing habits are challenged. The film shows itself for what it is and we protest that, 'it doesn't make sense!' By using the same techniques of other movies, such as cutting shots, filming from different viewpoints, and cliché, Lynch's film plays with the viewer's understanding of time, space and identity. Cliché is central to the interpretation of this film and goes some way to explaining its unsettling effect on the viewer. Perhaps the predominant cliché in the film is the revenge plot. This cliché is apparently resolved when Fred is caught and imprisoned for the murder of his wife on the basis of clear video evidence, but the cliché is subsequently undermined when the same man escapes to kill the evil mafia boss who is in fact responsible for the murder of his wife despite the evidence to the contrary. This revenge plot is connected with another major cliché surrounding the mystery of who really killed Fred's wife. This cliché of the murder mystery drives the entire film, yet by the end of the film the underlying issue is still largely unresolved with the protagonist undergoing yet another metamorphosis. We are asked to believe that the real killer was the evil mafia boss, even in the face of clear evidence to the contrary, but if Fred could metamorphose into somebody else to escape prison, why couldn't the mafia boss also transform himself into Fred? The power of metamorphosis runs throughout the film and serves to challenge and undermine the prevailing clichés. If the cliché of the traditional murder mystery is 'whodunit?' then *Lost Highway* turns that very question on its head. Even the cliché of the linear plot is undermined. The simple fact is that every film is commonly made of scenes that could have been recorded in almost any order and which are then arranged in an order chosen by the editor or director. The imposed linearity of the film is made abundantly clear to us, as are the gaps in time that are usually glossed over as we pass from scene to scene. Any apparent linearity is contrived, there is certainly no necessity to their order presented on screen. *Lost Highway* may be considered a primer in cliché where many more clichés could be unearthed in this analysis, including the prison escape, the struggle between good and evil, and the femme fatale.

If the camera never lies, it works by means of cliché, where one is forced to ask what is the nature of the truth it is telling? Nietzsche posed a very similar problem in the following aphorism, 'You may lie with your mouth, but the mouth you make as you do so you none the less tell the truth' (Nietzsche 1990b: 105). This is remarkably suited to the cinema where it could be rephrased as follows, 'The camera may lie with its images, but with the way it frames its images as its does so, it none the less reveals the truth of film making'. This is the critical dimension of Lynch's movies; they are beautiful representations that work by revealing their purely artificial nature. Lynch takes standard cinematic techniques such as montage, cutting shots and jumps in the time sequence, and shows these up for what they are. The presumed linearity of time and image are replaced by a highly mobile image. Cinema itself means movement, derived from the Greek, *kinemo*. This is what we might call the lost highway of information technology, the fact that linearity in time and movement is an invention, an imposition that becomes forgotten through habitual usage. In the hands of Lynch, the information technologies of film are used in all kinds of mobile, non-linear ways. His approach to film might be compared to the disruptive techniques of information warfare, scrambling lines of communication and undermining the mainstream propaganda machine.

### *Negri: the creative time of the multitude*

Like Nietzsche and Bergson, Antonio Negri also outlines a qualitative conception of time, but his re-evaluation gives it an important collective dimension. This collective conception of time exists outside the dictates of systems of social control. The project begins by defining productivity independently from mere ownership: '*Productive labour* is ... *not* founded as the relation with capital but on its own co-operative essence' (Negri 2003: 72, emphasis in original). Drawing upon a huge breadth of sources – including Marx, Nietzsche, Bergson, Spinoza, Heidegger, Foucault and Deleuze – Negri goes on to fashion a qualitative conception of time; which he terms *kairos*. *Kairos* is conceived as a time that is, in essence, collective and creative. Collective time is the origin of all forms of productivity. Negri claims that even times of collective struggle and sabotage are productive and should be seen as experiments in developing forms of higher quality work. It was labour that objected and resisted the mundane mind-destroying work of the factories and mills, which led to a re-evaluation of work by forcing the development of more efficient tools rather than a mere quantitative increase in burden. This is a continuation of a theme from Hardt and Negri's earlier book *Empire*. In this work, they identified the labour strikes and student rebellions of the 1960s as a path of resistance that was followed by capital in a general re-evaluation of the importance of lifestyles and leisure time to people, and to the new significance of intellectual and emotional labour. The multitude triggers a transvaluation of values. The rebellion of the multitude is a creative event, where the productive forces of the present open up to the future.

Negri has identified a key problem at the heart of the productive apparatus in this post-modern era, which he encapsulates in the following statement: 'The value produced by the multitude is immeasurable' (Negri 2003: 232). One does not have to be a follower of Marx to understand that measuring productivity is becoming increasingly problematic. Take for instance the recent concern for knowledge management, where corporations have been experimenting with ways of measuring intellectual assets, social capital and the value of tacit knowledge in general. A clear monetary measure may not be given to such 'assets', but their value to many organizations is obvious from their interest in knowledge management programmes and intellectual property rights. Many companies have now begun creating knowledge maps in an attempt to keep track of potentially valuable knowledge and skills, including Microsoft, Monsanto, Hoffman-Laroche, Health Canada and even the US Army. All sections of society are being confronted by the immeasurable nature of labour. A strike action being undertaken by the Association of University Teachers in the UK raised similar issues. The employers had penalized strikers by docking the pay of any days of work lost due to the action, as indeed is the norm in such situations. This would amount to the loss for each lecturer involved in the action of 1/365 of a year's salary for each day on strike. However, some academics raised the question, what if they make up for the lost work at a later time, as is very likely due to the nature of their academic work? Would they be recompensed if no loss in productivity could be proven? Also, if an academic were to work overtime at any later date would this be similarly accounted for? In fact, how can one measure the value of this kind of work by means of such a straightforward calculation? The measurement of labour-time is quite clearly contentious in such situations. When work includes so-called social skills, the ability to cooperate and to provide some kind of service, it becomes unclear what exactly falls outside the realm of production.

We have already looked at the role of information age nomads, such as the Linux software community, the Napster music community and the Zapatistas among others, who have all been pioneers of strategies to create a smooth space out of the world's communication networks. This argument also held that these groups should not be seen as parasitical on the corporate world because they play a fundamentally innovative and productive role, both in terms of the new products (e.g. Linux), new ideas (e.g. shareware, GNU copyleft) and new forms of human community. The argument that Negri puts forward goes much further than these relatively modest claims. Not only are cooperative communities, such as Linux and Napster, certainly not parasitical, it is in fact capital that is the parasite. Capital produces nothing. If it can be said to have created anything, it is a massive system of anti-production, a system of command and control designed for the purposes of exploiting the efforts and creative powers of others. Creativity comes directly from the cooperative efforts of the multitude. The creative aspect of *kairos* derives from the power of the imagination to create new relations; it is the 'creator of new being' (Negri 2003: 153). The imagination is not simply the result of idle fantasy, but is a creative expression embodied in language. Language, in fact, is the primary productive force in Negri's conception of cooperative time. Negri argues that, as a linguistic event, the act of creation must also be a communal event, which arises as a direct result of our cooperative efforts. Language is itself a productive force because it provides the basis for cooperation that is necessary to all forms of social organization and productive activity. Language, therefore, is not a private good, but is a tool held in common by us all:

> Language is not born and does not develop other than in the common and from the common. Nothing is produced that is not produced through the common: all commodities have become services, all services have become relations, all relations have become brains, and all brains form part of the common. Language is no longer only a form of expression; it is the only form of production of the human and its environment.
>
> (Negri 2003: 189)

Language is fundamental to post-industrial modes of production, both in the formation of networks of cooperation and in terms of the expression of innovation. Through language ideas can be generated and expressed, and then secondarily transformed into intellectual property. It is through language that services are provided, which create feelings of care, of pleasure, of excitement. Most crucially of all, language forms the basis of all cooperative human activity, and through it has emerged the world's cultures and diverse forms of subjectivity. This conception of language can also be drawn on to explain the central role of language in information warfare, especially in terms of the word virus, which has already been discussed in some detail in Chapter Five. According to Negri, at a basic level language is used as a tool, but it also acts in a far more complex way as a '*praxis* of truth' (Negri 2003: 180). Today, the means of production are not only concerned with the production of goods and services, but with the production of knowledge and truth. This biopolitical context directly concerns the circulation of information by the mass media and the production of science and technology. Taking responsibility for the truth means taking control of the means of production of truth, it is not enough merely to be informed by the dominant propaganda of the day. Language is essential to the creation of subjectivity and the networks of cooperation that serve as the basis for all forms of productive activity. It is biopolitical to the core, where biopolitics concerns the production of subjectivity, community and the truth itself.

Who owns this productive tool of language? It is certainly not restricted to any specific section of the population; it is everyone, it is the multitude and, most particularly, it is the poor. Continuing a theme from his earlier work, *Empire*, Negri believes the poor to be the basic productive force of society; and this is made absolutely clear when he describes the relationship between language and the poor: 'To give meaning to languages and to innovate in the cause of the circulation of meaning: these are the gifts that only poverty can bring' (Negri 2003: 203). Negri draws on the exemplar of St. Francis of Assisi both in *Empire* and his later work on *Time for Revolution*. St. Francis is the sign of poverty and his story illuminates the condition of the poor and their nearness to god. As the major productive force of the world, the poor are said to have god-like power; in the words of Hardt and Negri (2000: 157) 'only the poor has the ability to renew being… "The poor is god on earth"'. Their relationship to god and to the creative power of the world leads Negri to claim that the poor are the most exposed to the immeasurable (Negri 2003: 194). The condition of the poor is always a perilous one, where they may be said to live on the edge of being. They are the closest to death, in direct confrontation with the immeasurable. The life of the poor is lived as a common cooperative time. As a primary biopolitical force the life of the poor is immeasurable, and so is the value they generate.

## The time of information warfare: towards a biopolitical time

The argument of this chapter has contrasted a quantitative, measurable conception of time that is integral to systems of control and exploitation, with a number of qualitative conceptions, notably becoming, duration and *kairos*. All of these latter conceptions of time agree in their critique of a spatialized, measurable time; however, each has very different implications. Nietzsche's becoming and Bergson's duration each express time in fundamentally creative terms. However, only Negri's account of *kairos*, which is both a creative and collective conception of time, fully addresses the issues that are presented by information warfare. The techniques of information warfare are a clear recognition of a truly biopolitical society, where our communicative activities and our social relations have fallen almost completely under the sway of the dictates of business. Real time is the time of information warfare. The emergence of computer-mediated real time is directly related to innovations in systems of control that have computer technology at their heart. Information warfare is a practice that has emerged in part to exploit the power of these new technologies and in part to resist the same. Information warfare can thus be seen to be operating on two levels. On one level it acts as part of a system of real-time control and constitutes a system of command that has already been described in detail by writers such as Burroughs, Deleuze, Foucault and Negri. On another level, this kind of information warfare is, from its outset, a response to another more subversive form of information warfare, which is directed at these very systems of command and control. This second form of information warfare has to employ any resources that are available to it and is often low-tech in nature, owing to the limited resources of those who practice it. This subversive information warfare is aimed at the propaganda machine of the powerful, and at the growing systems of control that are being employed to regulate our lives and movement.

# 8 Conclusions, connections, lines of flight

> ...and ye shall know the truth, and the truth shall make you free.
>
> (CIA motto; also John 8:32)

> And I began to learn, gentlemen. Oh yes, one learns when one has to; one learns if one wants a way out; one learns relentlessly. One watches over oneself with a whip; one flays oneself at the slightest sign of resistance.
>
> (Kafka 1992: 195)

The above quotation comes from a short story by Franz Kafka called, *A Report to an Academy*. The speaker happens to be an ape that has learned to imitate humans and to speak the language of his captors. The ape is invited to explain his remarkable circumstances to the scientists and scholars of his day. This monologue is fascinating, not least because it explains that the ape had no innate desire to imitate men, but did so only as a means of escape, since he could not leave his cage by forcible means. The ape's becoming human can be described as a line of flight. Although he achieved freedom from his cage, he lost the freedom granted him by his animal nature and became entrapped within another cage, his adopted humanity and its social constraints. Since the Enlightenment, knowledge and learning has often been cited as a way out of submissive tutelage; it is the path to freedom. Kafka's absurd tale highlights an alternative dimension to knowledge and learning. Many comparisons can be drawn between the ape of Kafka's tale and the so-called 'flexible worker' that has emerged over recent years, responding to structural changes within information societies. The characteristics of a flexible worker have been described in terms of: adaptability to market developments, working unsocial hours, working from home, a commitment to life-long learning and a very high degree of self-management (Hielscher 2003, Huws 2003, Krings 2003). The freedoms include new job opportunities, being freed from the strictures of the traditional working day, and being able to organize your own leisure time. The disadvantages are a loss of previous forms of social protection, an increase in part-time and temporary work, an intensification of self-discipline and an increasing investment of one's own personality into the job. Learning is both a means of escape and a means of further entrapment. The emergence of different techniques for information warfare that have been outlined in this book have also been explained as lines of flight, and strategies for regulating and exploiting these lines.

If you have been the subject of identity theft or credit card fraud, if your computer has been infected with a virus, if you have been the target of a campaign of an advertising firm or political spin-doctor, if you have been the subject of malicious gossip and rumour, then you have been the subject of information warfare. You may not be interested in information warfare but information warfare is interested in you (to paraphrase Tolstoy). This book has outlined various strategies of control and resistance that are being developed to exploit the potential of information and communications technologies. Part I examined the organizational implications of information in term of its value as a resource and in terms of the evolution of nomadic strategies. Strategies for control have been developed for mapping and transforming knowledge into a territory, which can be regulated, subject to control and exploited as a commodity. These techniques include social innovations, such as intellectual property rights and knowledge management programmes, and technological innovations, such as computer-mediated networks of communication and real-time control systems. Several features were identified that make knowledge difficult to transform into a territory. For example, it is a non-rivalrous product and requires artificial barriers to prevent its spread. Also, it is difficult to alienate from the producer where brains and language are the means of production. Part I also explored alternative approaches to knowledge management and developed a theory of information nomadism. This theory took two distinctive forms, one of which was the development of a nomadic strategy that was derived largely from the work of T.E. Lawrence showing that nomadic strategy is ideally suited to communications-based organizations. Nomadic strategies exploit the potential of alternative communications networks to protect local cultures from the destructive effects of global capitalism. Nomadic strategy can be employed to further specific local political aims, where the strategy itself is not merely a means to an end but embodies a particular nomadic way of living. A series of cases were analysed, including the Zapatista rebels in Mexcio, who are attempting to preserve their indigenous culture, and the Slow Food Society, which began in Italy to protect its local traditions of agriculture and cookery against the invasion of the Fast Food industry. Nomadic strategies have also been seen in the sudden growth of online communities and user groups, such as Linux and Napster. These diverse organizations all use a strategy based around the nomadic principles of the pack, the idea, speed, communication, peace, friendship, leadership and calculation. The second form of information nomadism regards knowledge as a labyrinth rather than as a territory, and advocates the creation of an intellectual commons that could be an open resource for the benefit of all. This idea of the intellectual commons can be expressed in many different forms. In the realm of computing it can be found in the ideals of the Free Software Foundation, and in the realm of publication it exists in the ongoing development of an Open Public Licence, and MIT's Open Courseware. OPL may be the most remarkable social innovation since the legal embodiment of intellectual property rights. OPL is free property, which can be exchanged between people but not sold for money. This intellectual commons directly concerns the free flow of ideas, information and the preservation of local knowledge. This commons may be best expressed in the existence of language itself, which can be seen as an intellectual resource that is held in common by all; no one owns language.

Part II discussed the implications of treating information as a weapon. This argument outlined the history of information warfare and the armoury of techniques that have been developed for its practice. Information warfare has been analysed in terms of its object, its methods, and its specific weapons. The techniques of information warfare can be targeted against information resources and an individual or community's identity. Its methods include disruption, speed, propaganda, public relations and intelligence. A range of high technology weapons have been developed for information warfare, such as the virus, email bombs, virtual sit-ins and many other kinds of Internet and computer-based attacks. Low technology weapons can also be employed for the purposes of information warfare, including rumours, slogans, confidence tricks, and propagandist pamphlets and leaflets. The virus has been identified as the perfect weapon for information warfare. The virus was also shown to be a key metaphor for understanding the transmission of information, ideas and language within the global communication networks and the ways in which organizations are adapting to these networks.

Part III of the book looked specifically at changing power-relations, of which the techniques of information warfare may be seen as a manifestation. This argument examined the idea that we are moving away from disciplinary societies to societies based on free-floating mechanisms of control. In disciplinary societies the predominant diagram of power could be described in terms of the Panopticon, which operated within fixed sites and a visual architecture of surveillance. During the twentieth century, control societies adapted free floating networks that operate by connecting nodes, expanding networks and integrating all aspects of the social domain into their webs of control. The net (or web) metaphor is illuminating in at least two contrasting senses, on the one hand a net is an interconnected series of nodes, whose power derives from its openness and the proliferation of its lateral connections, but on the other hand, a net also serves as a mechanism of capture and may be seen as a kind of trap.[1] These networks operate by modulating the interactions between nodes on a principle of homeostasis, rather than by imposing a fixed norm or moulding as was done in disciplinary institutions. Different forms of social control can be associated with particular innovations in technologies for the measurement of time. A technological threshold has been crossed that distinguishes the kinds of power relations that predominated under the disciplinary diagram of power from post-modern networks of control. Real-time techniques of control are now serving to enhance the cybernetic interactivity of the network society, by such means as corporate supply chains, loyalty cards that capture consumer loyalty, and the real-time investment decisions of speculative capital. Real-time systems of control are a key tool for the invasion and transformation of new territories for exploitation, whether this is conceived in terms of a business expanding into new markets or an army expanding its military operations. Real-time is the time of information warfare. Information resources and communication networks are a fundamental part of the biopolitical apparatus of control. This biopolitical apparatus has grown both in terms of its intensive and extensive influence. It has grown extensively in terms of the geographical spread of international and multinational markets and corporations, and intensively in terms of its increasing infiltration into the realms of so-called private life, which involves the exploitation of social skills, ideas, and communication.

But why focus on subversives like Burroughs or the Zapatistas? For one thing, these people have been genuine pioneers in the techniques of information warfare, and this has been recognized not just by *avant garde* thinkers like Gilles Deleuze, but also within mainstream think tanks like the RAND Corporation. Deleuze observed that society is defined to a large extent by its sites of struggle, and by what he termed its 'lines of flight', or the new social movements and ways of living that have escaped the old struggles. Innovations in the way we live and work often follow these lines of flight. For example, the rise in the value of informational and immaterial labour and the recognition of the importance of so-called 'lifestyles' and leisure pursuits followed the student protests and factory standstills of the 1960s (Hardt and Negri 2000). Although few organizations have adopted a nomadic strategy with regards to the use of information technology, they have been mimicking many of its innovations. The principles of speed and communication that are essential to nomadic strategy have also been cornerstones of the development of international business and the circulation of global capital. Business has had to adapt to absorb what runs outside its reach, but every new innovation by capitalism is met with further escape attempts by the nomadic war machine. For example, there has been an explosion of copying and piracy in the music industry and the pharmaceutical industry amongst many others, and a proliferation of strategies to escape copyright laws. Napster did not directly infringe the copyright legislation of time because it did not do any of the copying itself, but merely facilitated file sharing between millions of different computer users. Its weakness was that it still held a central database of Napster users who were infringing copyright and, after legal action by the RIAA, this was disconnected. Since the decline of Napster it has been followed by other more resilient file sharing communities such as Kazaa, which is located outside the US on the Pacific island of Vanuata and lies beyond the immediate reach of the US legal system. Gnutella is another file-sharing programme that has attempted to overcome the weaknesses of Napster because it is a purely distributed network that has no central point of control or coordination that can be closed down. The music industry has taken to the online distribution of its products after the initial shock of Napster, in an attempt to match some of the services offered by its illegitimate counterparts.

This is a book on strategy, which has described a range of strategies for the purposes of capture and control and contrasted these with strategies for the purposes of flight and escape. It has described the nomadic war machine as a fundamentally creative force, a way of living that is a strategy in its own right. It can be seen as a call to arms, to create and to protect the commons. This is by no means a call for violence. As Deleuze and Guattari (1988) have observed, the nomadic war machine is more concerned with flight, innovation and art. Nomadic strategy is aimed at avoiding battle, it is not distinguished by violence but by the way in which it occupies and creates a smooth space within which the nomads can move. In fact, it is not the nomad but the state and its agents that resort to violence to contain the nomad, with a sedentary morality to provide the necessary justification (following Joseph Conrad, one might better call it the mock morality of the Company). Nomadic strategy only becomes violent when it hits a wall, and its movement is impeded. The nomad always prefers flight. It is the nomads way of maintaining their territory, occupying it by means of speed, legwork, by moving across space rather than by fencing it off. Nomadic strategy and violence partially converge in guerrilla tactics. Again, the guerrilla tries to avoid direct confrontation, knowing that battles will generally favour the forces of the state or the Company. Guerrillas use speed, evasion, ideas and communications networks to enhance their meagre resources. The guerrilla attempts to make it impossible for occupying forces to hold a territory, disrupting the striated networks of communication and transportation. In many respects the new communications technologies are ideal mechanisms for guerrilla strategy, and it is no surprise that groups such as the Zapatistas have attempted to exploit these. Such technologies afford guerrilla groups a new freedom of movement to spread their own ideas and propaganda independently from the mass media, and to interrupt the information systems of the elites. This freedom allows them to turn new technologies of communication, such as the Internet into a smooth space in which they can circulate relatively free from interference.

In the information economy, the distinction between tools and weapons becomes problematic. They both share the same line of flight, where the kinds of information warfare being conducted by the military are mirrored by techniques used in business, which are becoming generalized throughout many levels of society. The same techniques of propaganda and marketing, hacking, disruption and deception are being used by both military and civilian groups, by the state, by corporations and by protest groups. The civilian and the soldier are increasingly relying on information as a means of work and warfare, and both are vulnerable to attacks on their information systems and information resources. Like any other tool, an information system can become a weapon depending on exactly how it is used. If it is used in an introceptive manner, as a basis for production, then it becomes a tool of work that can be exploited and from which surplus value can be extracted. If it is used in a projective manner, in order to disrupt an opponent's movements or interrupt their way of life, then it becomes a weapon. A weapon is certainly useful, but it is not a part of the process of production and its use cannot be directly exploited as creating value. It can of course be used to steal or destroy things that have already been created. The weapon is an instrument for undoing. This finds its ideal expression in the martial arts where the weapon is not only aimed outwardly against an opponent but is part of an ascetic practice centred on the martial artist. The martial artist sets out to challenge their ego and unravel it through constant practice and the ascetic demands of the art. It is the perfection of the nomadic war machine, where the practice of the art does away with the need for direct combat. Violence is subordinated to a 'way', which views the weapon as a provisory means to the perfection of its bearer. Can such a creature be found in the realm of information warfare? One might look to the masters of intelligence and the guardians of information in the CIA or IBM perhaps? But it would be highly contentious to claim that the employees of these monolithic organizations have managed to transform their work into a kind of martial art. Deleuze and Guattari (1988: 403) have observed that these organizations do provide an interesting point of departure, as a 'somber caricature' of the warrior. The closest approximation to the warrior type in information warfare can be found in the works of William S. Burroughs. In 1971, Burroughs wrote an article entitled *The Electronic Revolution*, which was perhaps the first manifesto that advocated the use of communications technologies for information warfare (Burroughs and Odier 1989). In this article he proposed that people use tape recorders to produce their own media, both to disrupt the messages of the mass media and to take the means of communication into their own hands. He outlined four specific techniques including: (i) spreading rumours, (ii) discrediting the speeches of authorities by interweaving recordings with shocking and unpleasant sounds and then playing them back in public places, (iii) using cut-ups of screams and gun shots in public places to escalate riots, and (iv) cutting up the messages of the mass media to scramble and nullify their lines of association. This overall strategy is framed by the following strategic principles,

> History shows that when a system of government is overthrown by force a system in many respects similar will take place. On the other hand he who does not resist force that enslaves and exterminates will be enslaved and exterminated. For the revolution to effect basic changes in the existing conditions three tactics are required: 1. Disrupt. 2. Attack. 3. Disappear. Look away. Ignore. Forget.
>
> (Burroughs and Odier 1989: 101)

Burroughs defined the role of the citizen as a guerrilla fighter who must actively struggle for democracy, which cannot exist in the first place without the means of communication being brought under the citizen's direct control. Burroughs conceived of this citizen fighter as akin to the Zen warrior. A key part of the discipline of the warrior is the struggle to overcome the ego or word virus, battling against the forms of addiction that have become the basis of the control society. In Burroughs' terms, we have nothing to lose except our prerecordings. From birth we are continually infected with the prerecordings of advertising firms and political spin-doctors. He was also the first to see the importance of the virus to mechanisms of social control, and his work is indispensable for understanding the virus as the paradigmatic form of information warfare. The virus has gained new prominence as a tool and a weapon with the current revolutions in our communications systems and biotechnology. The theorists of biopower and the control society have all acknowledged the importance of Burroughs' work as an influence on their own thinking, including Michel Foucault, Gilles Deleuze, and Antonio Negri.[2] The literary experiments of Burroughs should not be seen as being confined to the realm of fiction, because they also provide a practical manual of techniques for information warfare. Equally as important, this work is an expression of an ingenious linguistic imagination, which has created a language that is simply impossible to control, subversive to the core.

One of the most notorious champions of information warfare in the 1990s was Kevin Mitnick, whose work and activities have been discussed earlier in this book. Mitnick's approach also emphasizes the importance of informal communications as a way of gaining access to restricted information, a practice that he terms 'social engineering'. His work highlights the fact that the weakest link in the security of any organization is often its own employees. Mitnick's work can be drawn upon to show that the very same characteristics that make up the building blocks of the service economy, such as trust, cooperation and a willingness to communicate, can be exploited by the hacker or social engineer. The social engineer uses his or her social skills to establish a relationship with the target, attempting to gain their trust and mislead them. No matter how good the technology, the human factor can always be exploited. Authors such as Mitnick and Denning show that hackers and social engineers take advantage of our gregarious nature. These authors emphasize the illegitimate nature of these kinds of activities, and they provide very useful guidelines for protecting one's own information resources from attack. However, this work often neglects similarities between legitimate and illegitimate approaches to information warfare. For example, Mitnick explained the use of reciprocation as a technique, where a 'free' gift is offered, which is used to influence the target to return the favour. Mitnick warns against this kind of thing, referring to the case of the Hare Krishna cult, which has used this technique to increase the likelihood of reciprocal donations to their organization. He neglects to mention that the offer of a 'free' toy or voucher is also a standard marketing gimmick, but surely no less a form of social engineering. A similar confidence trick that Mitnick describes for attaining the participation and cooperation of a target is presenting them with a 'special offer' that is supposed to be in short supply or available for a limited period of time. Many of these simple techniques of social engineering and information warfare are almost precisely the same as the supposedly legitimate practices of marketing or public relations, a fact which would seem to put the latter's legitimacy into question. Mitnick's case could be seen as a classic example of poacher turned gamekeeper. After doing time in prison, Mitnick is a reformed man and he now recommends that the employee should be motivated 'to want to *chip* in and do his part to protect the organization's information assets' (Mitnick 2002: 250). Whereas once he used his social and technological skills to break into the information systems of big business, he now uses these same skills in the service of these businesses. Although his work provides an excellent introduction to information warfare and good advice for the individual or the corporation on how to protect against illegal attacks, this work does not admit that we are subjected to precisely the same kinds of attacks from legitimate organizations.

Denning (1999) also concedes that advertising is a possible form of information warfare, but argues that this is only true if the advertisement positively misleads its audience. The work of Burroughs, in contrast, does not make these distinctions and recognizes such attacks for what they are, regardless of their perceived legitimacy. Burroughs proposed that we should raise our awareness and protect ourselves from forms of information warfare independently from our moral prejudices, which may well be product of previous control campaigns. The work of Subcomandante Marcos of the Zapatistas serves as a more consistent exemplar of information warfare. The Zapatista slogan of 'our word is our weapon' shows many similarities with Burroughs' approach, as does their active engagement with alternative channels of communication. Like Burroughs, Marcos also grapples with issues concerning the creation of an identity that can be maintained independently from the apparatus of command

and control (Marcos 2001). Perhaps we should not be looking to an individual, but to a community or a network as the true champion of information warfare. Hardt and Negri (2000) coined the term 'the multitude' to describe such a network. For Burroughs, it was the underground organization that he called the Johnson Family. In the case of Subcomandante Marcos, it is the interconnecting web of human rights groups, unions, protestors and armed militia that make up the Zapatistas who form a distributed network to 'name the nameless' (Marcos 2001). T.E. Lawrence defined the driving force behind his strategy in terms of the ideals of freedom and the nomadic life. This movement was built up from an undisciplined mix of local tribes and British soldiers, all of whom drew upon the limited resources of camels, tents, rifles, explosives and bully beef. These factors allowed the creation of a highly effective strategy that emphasized the power of speed and communication, and which could triumph within the harsh discipline imposed upon it in the desert: 'suppose we were . . . an influence, an idea, a thing intangible, invulnerable, without front or back, drifting about like a gas' (Lawrence 1962: 198). Under certain material conditions the pack is a more appropriate organizing principle than the well-disciplined corps. Our communications and interactions need not be guided by ethics committees comprising well-qualified wise man, where packs of self-organized user groups can do a far more effective job (Deleuze 1995). Indeed, this has been the basis of groups such as the Linux user community and the Slow Food Society, among many others. The champion of information warfare is an assemblage of ideas, packs and communications networks. Within information societies this assemblage has found its ideal expression in information nomadism, and the creation of distributed networks that afford free movement across a smooth space.

# Notes

## 2 Strategies for the Information Age: nomadic wisdom and guerrilla warfare

1 It may be countered that wealth also allows for a freedom of movement that is denied to the impoverished. This is certainly a truism. However, the quality of this form of movement is of a very different character, being more associated with merely quantitative distance travelled than the intensity of nomadic speed. Take for instance, the package holiday as an example of the former where the movement is characterized by moving from more or less identical cosmopolitan locations, cities, hotels, and so on. One is constantly looked after and entertained by the host company and its representatives. In contrast, the quality of movement that nomadism fosters is more ascetic in nature and requires freedom from such attachments.

2 The term 'GNU' is just a self-referential acronym meaning 'GNU's Not Unix'.

## 3 Mapping knowledge: the new cartographers of organization

1 An exception to this is the work of Flower and Heath (1993). These authors examine the disciplinary effects of the Human Genome Project. In particular, they discuss how 'micro-anatomo politics' is being becoming a new tool in the control of populations, or what Foucault termed 'biopower'. They also note that the genetic map has been primarily a map of defects.

2 The film *Gattaca* gives a powerful portrayal of just such a vision of such a future.

## 4 Organization, society and information warfare

1 I would like to acknowledge the website 'Pampleteers and Web Sites' by Dan Bricklin at www.bricklin.com/pamphleteers.htm as the source of some of these ideas.

2 The regulation of access to communications media is a consideration for governments of all countries, whether democratic or not. The US government, for example, has left the Internet relatively free of regulation but closely regulates access to satellite imagery. In contrast, the Chinese government has been experiencing great problems in regulating its people's access to the Internet, which has been embodied in law Order No. 195, 'Interim Regulations on International Interconnection of Computer Information Networks in the PRC'.

3 I stress the word 'relative' here since control clearly has been exercised, but usually through the more subtle techniques of propaganda and public relations.

4 Strictly speaking, David Ronfeldt is a RAND analyst, and John Arquilla is a

defence analyst at the Naval Postgraduate School, however much of their joint work is published by the RAND Corporation.

5 Divine sanction is often used by both sides in a conflict oblivious to the irony of their positions.

6 A short anecdote from my own experience can serve to illustrate how censorship works in the market system. A few years ago I went to see the great American novelist Gore Vidal speak at the Amnesty International lectures at Oxford, UK. As an aside during his lecture he told the audience about a recent documentary that he had made about Washington DC for the BBC. I had seen this documentary and had thought it both informative, well balanced and critical. Not so, Mr Vidal told us, in fact a large part of the documentary he had made covered the ghettos of Washington and its endemic racism. This however had been edited from the final product, since some of the intended customers were US media companies, and the BBC was not keen to alienate this important market. So the propaganda model may be in operation on many levels even in the absence of a conscious decision to be propagandist.

7 This assemblage of culture, geography and resistance to the state has been termed 'the nomadic war machine' by Deleuze and Guattari (1988), which beautifully explains the creativity and power and this kind of organization.

8 It may be noted that there is a religious flavour to Virilio's analysis of the accident, and he himself says that strategists should be reading the Gospel of St John rather than Sun Tzu.

9 For an excellent explanation of this kind of blowback, with particular reference to its historical context, the reader is directed to Kolko (2002) for a concise analysis of the ongoing crisis in Afghanistan and the Middle East.

10 'Sup' is one of Subcomandante Marcos' many nicknames.

## 8 Conclusions, connections, lines of flight

1 My thanks to Dawn Rimmer for bringing my attention to the contrasting implications of the net or web.

2 Many of the theorists of post-modern forms of control attended a lecture given by William Burroughs in the US in 1975, including Deleuze, Foucault and Lyotard.

# Bibliography


Adamson, D. (2000) Science and philosophy: two sides of the absolute, *PLI Warwick Journal of Philosophy*, 9, pp. 53–85.

Adelstein, T. (2001) *Is Microsoft Engaged in an Information War Against Open Source or IBM?* Online. Available http://www.consultingtimes.com.

Alderman, J. (2001) Free for all, *Guardian's Weekend* (8 April), pp. 50–55.

Adorno, T. (1991) The schema of mass culture, in *The Culture Industry: Selected Essays on Mass Culture*, London: Routledge, pp. 53–84.

Alvesson, M. and Karreman, D. (2001) Odd couple: making sense of the curious concept of knowledge management, *Journal of Management Studies*, 38, pp. 995–1018.

Arquilla, J. and Ronfeldt, D. (eds) (1997) *In Athena's Camp: Preparing for Conflict in the Information age*, Washington: RAND National Defense Research Institute.

Arquilla, J. and Ronfeldt, D. (1997a) A new epoch – and spectrum – of conflict, in Arquilla, J. and Ronfeldt, D. (eds) *In Athena's Camp: Preparing for Conflict in the Information age*, Washington: RAND National Defense Research Institute, pp. 1–20.

Arquilla, J. and Ronfeldt, D. (1997b) Cyberwar is coming, in Arquilla, J. and Ronfeldt, D. (eds) *In Athena's Camp: Preparing for Conflict in the Information age*, Washington: RAND National Defense Research Institute, pp. 23–60.

Arquilla, J. and Ronfeldt, D. (1997c) Information, power, and grand strategy: in Athena's camp – Section 1, in Arquilla, J. and Ronfeldt, D. (eds) *In Athena's Camp: Preparing for Conflict in the Information age*, Washington: RAND National Defense Research Institute, pp. 141–171.

Arquilla, J. and Ronfeldt, D. (1997d) The advent of netwar, in Arquilla, J. and Ronfeldt, D. (eds) *In Athena's Camp: Preparing for Conflict in the Information age*, Washington: RAND National Defense Research Institute, pp. 275–294.

Arquilla, J. and Ronfeldt, D. (1997e) Information, power, and grand strategy: in Athena's Camp – Section 2, in Arquilla, J. and Ronfeldt, D. (eds) *In Athena's Camp: Preparing for Conflict in the Information age*, Washington: RAND National Defense Research Institute, pp. 417–438.

Arquilla, J. and Ronfeldt, D. (1997f) Looking ahead: preparing for information-age conflict, in Arquilla, J. and Ronfeldt, D. (eds) *In Athena's Camp: Preparing for Conflict in the Information age*, Washington: RAND National Defense Research Institute, pp. 439–501.

Arrow, K. (1984) *The Economics of Information*, Oxford: Blackwell.
Baran, P. (1964) Introduction to distributed communications, *RAND Memorandum RM-3420-PR*, Washington, RAND.
Barbrook, R. (2001) *The Napsterisation of Everything: A Review of John Alderman, Sonic Boom: Napster, P2P and the Battle for the Future of Music*, Fourth Estate. Online. Available, The Cyber-Society-Live mailing list.
Barr, J. (2000) *Opinion: Napster, Gnutella, and Internet guerrillas*, Online. Available http://www.cnn.com/2000/tech/computing/05/12/mp3.guerillas.idg/index.html.
Barthes, R. (1977) *Image, Music, Text*, London: Fontana Press.
Barthes, R. (1979) *A Lover's Discourse: fragments*, London: Penguin.
Barthes, R. (1990) *A Lover's Discourse: Fragments*, London: Penguin.
Bechmann, G., Krings, B. and Rader, R. (eds) (2003) *Across the Divide: Work Exclusion: Work, Organization and Social Exclusion in the European Information Society*, Berlin, edition-sigma.
Beer, S. (1979) *The Heart of the Enterprise*, Chichester: Wiley.
Bergson, H. (1910) *Time and Free Will: An Essay on the Immediate Data of Consciousness*, London: George Allen and Unwin.
Bergson, H. (1912) *An Introduction to Metaphysics*, translated by T.E. Hulme, Indianapolis: Hackett Publishing.
Bergson, H. (1999) *Duration and Simultaneity*, Manchester: Clinamen Press.
Bernays, E. (1923) *Crystallizing Public Opinion*, New York: Liveright.
Bernays, E. (1933) *Propaganda*, New York: Liveright.
Bernays, E. (1952) *Public Relations*, University of Oklahoma Press.
Bogue, R. (1999) Art and territory, in Buchanan, I. (ed.) *A Deleuzian Century*, Durham: Duke University Press, pp. 85–102.
Bougen, P. and Young, J. (2000) Organizing and regulating as rhizomatic lines: bank fraud and auditing, *Organization*, 7, pp. 403–426.
Boje, D., Gephart, R. and Thatchenkery, T. (eds) (1996) *Postmodern Management and Organizational Theory*, Thousand Oaks: Sage.
Burrell, G. (1992) Back to the future: time and organization, in Reed, M. and Hughes, M. (eds) *Rethinking Organization*, London: Sage, pp. 165–183.
Burrell, G. (1997) *Pandemonium: Towards a Retro-Organization Theory*, London: Sage.
Burroughs, W.S. (1968) *The Ticket that Exploded*, London: Flamingo.
Burroughs, W.S. (1986) *The Adding Machine: Selected Essays*, New York: Arcade Publishing.
Burroughs, W.S. (1995) *My Education: A Book of Dreams*, London: Picador.
Burroughs, W.S. and Odier, D. (1989) *The Job: Interviews with William S. Burroughs*, London: Penguin.
Calasso, R. (1995) *The Ruin of Kasch*, London: Vintage.
Calvino, I. (1997) *The Literature Machine*. London: Vintage.
Canetti, E. (1973) *Crowds and Power*, London: Penguin.
Carter, P. and N. Jackson (2000) *Rethinking Organisation Behaviour*, London: Pearson Education.
Castells, M. (1996) *The Rise of the Network Society*, Oxford: Blackwells.
Castells, M. (1997) *The Power of Identity*, Oxford: Blackwell.
Castells, M. (2001) *The Internet Galaxy: Reflections on the Internet, Business and Society*, Oxford: Oxford University Press.
Chatwin, B. (1987) *The Songlines*, London: Vintage.

Chomsky, N. (1992) *Deterring Democracy*, London: Vintage.

Chomsky, N. (1998) Propaganda and control of the public mind, in McChesney, R., Wood, E. and Foster, J. (eds) *Capitalism and the Information Age: The Political Economy of the Global Communication Revolution*, New York: Monthly Review Press, pp. 179–189.

Chomsky, N. (1999) *Profit over People: Neoliberalism and the Global Order*, London: Seven Stories Press.

Colville, I., Waterman, R. and Weick, K. (1999) Organizing and the search for excellence: making sense of the times in theory and practice, *Organization*, 6, pp. 129–148.

Colby, W. (1992) Privatizing intelligence, in Sigurdson J. and Tagerud, Y. (eds) *The Intelligent Corporation: The Privatization of Intelligence*, London: Taylor Graham, pp. 17–22.

Cook, S. and Brown, J. (1999) Bridging epistemologies: the generative dance between organizational knowledge and organizational knowing, *Organization Science*, 10, pp. 381–400.

Cummings, S. (1993) Brief case: the first strategists, *Long Range Planning*, 26, pp. 133–135.

Davenport, T. (1995) Some principles of knowledge management, *Business and Strategy*, 7, pp. 34–41.

Davenport, T. and Prusak, L. (1998) *Working Knowledge: How Organizations Manage What They Know*, Boston, MA: Harvard Business School.

Davenport, T., De Long, D. and Beers, M. (1998) Successful knowledge management projects, *Sloan Management Review*, 39, pp. 43–57.

de Cock, C., Fitchett, J. and Farr, M. (2001) Myths of the near future: advertising the new economy, *Ephemera: Critical Dialogues on Organization*, 1, pp. 201–228, Online. Available http://www.emphemeraweb.org/.

de Landa, M. (1991) *War in the Age of Intelligent Machines*, New York: Zone Books.

de Landa, M. (1997) *A Thousand Years of Non-Linear History*, New York: Swerve Editions.

Der Derian, J. (2002) Preface of Virilio, P. *The Information Bomb*, London: Verso, pp. vii–xvi.

Dorril, S. (2000) *MI6: Fifty Year of Special Operations*, London: Fourth Estate.

Deleuze, G. (1988) *Foucault*, London: Athlone.

Deleuze, G. (1983) *Nietzsche and Philosophy*, London: Athlone.

Deleuze, G. (1990) *The Logic of Sense*, New York: Columbia University Press.

Deleuze, G. (1991) *Bergsonism*, New York: Zone Books.

Deleuze, G. (1992) *Cinema 1: the Movement Image*, London: Athlone.

Deleuze, G. (1995) *Negotiations*, New York: Columbia University Press.

Deleuze, G. (1998) *Essays Clinical and Critical*, London: Verso.

Deleuze, G. and Guattari, F. (1983) *Anti-Oedipus: Capitalism and Schizophrenia*, London: Athlone.

Deleuze, G. and Guattari, F. (1988) *A Thousand Plateaus: Capitalism and Schizophrenia II*, London: Athlone.

Denning, D. (1999) *Information Warfare and Security*, Indianapolis: Addison-Wesley.

Eco, U. (1977) Striking at the heart of the state, *Apocalypse Postponed*, Indiana University Press, pp. 177–181.

Eco, U. (1984) *Reflections on the Name of the Rose*, London: Martin Secker and Warburg.

Eden, C., Jones, S. and Sims, D. (1979) *Thinking in Organizations*, London: Macmillan.

Ellul, J. (1965) *Propaganda: The Formation of Men's Attitudes*, New York: Vintage Books.

Elmer, G. (1999) Diagrams, maps and markets: the technological matrix of geographical information systems, *Space and Culture*, 3, pp. 40–60.

Flower, M. and Heath, D. (1993) Micro-anatomo politics: mapping the human genome project, in *Culture, Medicine and Psychiatry*, 17, pp. 27–41.

Ford, H. (1924) *My Life and Work*, London: Heinemann.

Foucault, M. (1971) *Madness and Civilization*, London: Routledge.

Foucault, M. (1977) *Discipline and Punish: The Birth of the Prison*, London: Penguin.

Foucault, M. (1981) *History of Sexuality: Volume 1*, London: Penguin.

Foucault, M. (1982) Afterforward: the subject and power, in Dreyfus, H. and Rabinow, P. (eds) *Michel Foucault: Beyond Structuralism and Hermeneutics*, Hemel Hempstead: Harvester Press, pp. 208–226.

Foucault, M. (1988) *Michel Foucault: Politics, Philosophy, Culture*, in Kritzman, L.D. (ed.), London: Routledge.

Foucault, M. (2000a) About the concept of the 'dangerous individual' in nineteenth-century legal psychiatry, in *Power: Essential Works of Michel Foucault 1954–1984*, London: Penguin, pp. 176–200.

Foucault, M. (2000b) 'Omnes et Singulatim': toward a critique of political reason, in *Power: Essential Works of Michel Foucault 1954–1984*, London: Penguin, pp. 298–325.

French, G. (2000) *Shunning the Frumious Bandersnatch: Current Literature on Information Warfare and Deterrence*, The Terrorism Research Center, www.terrorsm.com/analysis/iw-deterrence.shtml.

Fuchs, G. (2003) Working in the knowledge economy – the case of multimedia production, in Bechmann, G., Krings, B. and Rader, R. (eds) *Across the Divide: Work Exclusion: Work, Organization and Social Exclusion in the European Information Society*, Berlin, edition-sigma, pp. 217–234.

Fuller, S. (2001) A critical guide to knowledge society newspeak: or, how not to take the great leap backward, *Current Sociology*, 49, pp. 177–201.

Gray, J. (2003) *Straw Dogs: Thoughts on Humans and Other Animals*, London: Granta.

Guevara, C. (1998) *Guerrilla Warfare*, Lincoln: University of Nebraska Press.

Hammer, M. (1990) Reengineering work: don't automate, obliterate, *Harvard Business Review*, 68, pp. 104–112.

Hancock, P. and Tyler, M. (2000) 'The look of love': gender and the organization of aesthetics, in Hassard, J., Holiday, R. and Wilmott, H. (eds) *Body and Organization*, London: Sage, pp. 108–129.

Hardt, M. and Negri, A. (2000) *Empire*, Cambridge: Harvard University Press.

Harraway, D. (1997) *Modest_Witness@Second_Millenium. FemaleMan$^{c}$_Meets_OncoMouse$^{TM}$*, London: Routledge.

Hayles, N.K. (1999) *How We Became Posthuman: Virtual Bodies in Cybernetics, Literature, and Informatics*, Chicago: The University of Chicago Press.

Herman E. (1998) The propaganda model revisited, in McChesney, R., Wood,

E. and Foster, J. (1998) *Capitalism and the Information Age: The Political Economy of the Global Communication Revolution*, New York: Monthly Review Press, pp. 191–205.

Herman, E. and Chomksy, N. (1988) *Manufacturing Consent: The Political Economy of the Mass Media*, New York: Pantheon Books.

Hielscher, V. (2003) Flexible work and work life balance – potential or contradiction?, in Bechmann, G., Krings, B. and Rader, R. (eds) *Across the Divide: Work Exclusion: Work, Organization and Social Exclusion in the European Information Society*, Berlin, edition-sigma, pp. 161–172.

Hill, C. (1991) *The English Bible and the Seventeenth-Century Revolution*, London: Penguin.

Himanen, P. (2001) *The Hacker Ethic and the Spirit of the Information Age*, London: Secker and Warburg.

Ho, T.H. and Weigelt, K. (1997) Game theory and strategic thinking, in de Wit, B. and Meyer, R. (eds) *Strategy: Process, Content and Context*, London: International Thompson Publishing, pp. 103–119.

Hull, R. (1999) Actor network and conduct: the discipline and practices of knowledge management, *Organization*, 6, pp. 405–428.

Huws, U. (2003) The restructuring of employment in the information society and its implications for social protection, in Bechmann, G., Krings, B. and Rader, R. (eds) *Across the Divide: Work Exclusion: Work, Organization and Social Exclusion in the European Information Society*, Berlin, edition-sigma, pp. 139–153.

Illich, I and Sanders, B. (1989) *The Alphabetization of the Popular Mind*, London: Penguin.

Jaskolski, H. (1997) *The Labyrinth: Symbol of Fear, Rebirth and Liberation.* Boston, MA: Shambhala Publications.

Jermier, J., Knights, D. and Nord, W. (1994) *Resistance and Power in Organizations*, London: Routledge.

Kafka, F. (1992) *The Transformation ('Metamorphosis') and Other Stories*, London: Penguin.

Kasaras, K. (2002) Music in the age of free distribution: MP3 and society, *First Monday*, 7(1), pp. 1–18, Online. Available http://*www.firstmonday.org/issues/issue7_1/Kasaras*.

King, G. (1996) *Mapping Reality: An Exploration of Cultural Cartographies*, London: Macmillan.

Klein, N. (2000) *No Logo*, London: Flamingo.

Kofman, S. (1983) *Nietzsche and Metaphor*, London: Athlone.

Kolko, G. (2002) *Another Century of War?*, New York: The New Press.

Korzybski, A. (1958) *Science and Sanity.* 4th edn, Massachusetts: Colonial Press.

Krings, B.-J. (2003) Hen or egg? – The relationship between IC-technologies and social exclusion, in Bechmann, G., Krings, B. and Rader, R. (eds) *Across the Divide: Work Exclusion: Work, Organization and Social Exclusion in the European Information Society*, Berlin, edition-sigma, pp. 123–138.

Laqueur, W. (1977) *Guerrilla: A Historical and Critical Study*, London: Weidenfeld and Nicolson.

La Nuez, D. and Jermier, J. (1994) Sabotage by managers and technocrats: neglected patterns of resistance at work, in Jermier, J., Knights, D. and Nord, W. (eds) *Resistance and Power*, London: Routledge, pp. 219–251.

Lawrence, T.E. (1962) *Seven Pillars of Wisdom*, London: Penguin.

Lenin, V.I. (1964) On slogans, in *Collected Works*, London: Lawrence and Wishart, 25, pp. 183–190.

Lyotard, J.-F. (1984) *The Postmodern Condition: A Report on Knowledge*, Manchester: Manchester University Press.

Marcos, S. (2001) *Our Word Is Our Weapon: Selected Writings*, Ponce de Leon, J. (ed.), London: Serpent's Tail.

Margulis, L. and Sagan, D. (1986) *Microcosmos: Four Billion Years of Mircrobial Evolution*, London: University of California Press.

Margulis, L. and Sagan, D. (1997) *Slanted Truths: Essays on Gaia, Symbiosis and Evolution*, New York: Springer Verlag.

Marshall, E. (1997) Secretiveness found widespread in life sciences *Science*, 276, pp. 524.

Marx, G. (1948) Groucho Marx to Warner Brothers, in Kermode, F. and Kermode, A. (eds) *The Oxford Book of Letters*, Oxford: Oxford University Press.

Marx, K. (1976) *Capital – Volume 1*, London: Penguin.

Marx, K. (1934) *The Eighteenth Brumaire of Louis Bonaparte*, Moscow: Progress Publishers.

Marx, K. and Engels, F. (1998) *The Communist Manifesto: A Modern Edition*, London: Verso.

May, C. (2001) Unacceptable costs: the consequences of making knowledge property in a global society, *Global Society*, 16, pp. 123–144.

May, C. (2002) Trouble in e-topia: knowledge as intellectual property, *Urban Studies*, 39, pp. 1037–1049.

McChesney, R., Wood, E. and Foster, J. (1998) *Capitalism and the Information Age: The Political Economy of the Global Communication Revolution*, New York: Monthly Review Press.

McKinlay, A. (2002) The limits of knowledge management, *New Technology, Work and Employment*, 17, pp. 76–88.

McKinlay, A. and Starkey, K. (eds) (1998) *Foucault, Management and Organization Theory*, London: Sage.

McLuhan, M. (1964) *Understanding the Media: The Extensions of Man*, London: Routledge.

Mims, C. and White, D. (1984) *Viral Pathogenesis and Immunology*, Oxford: Blackwell Scientific Publication.

Minns, E. (1913) *Scythians and Greeks*, Cambridge: Cambridge University Press.

Mitnick, K. (2002) *The Art of Deception: Controlling the Human Element of Security*, Indianapolis: Wiley.

Moody, G. (2001) *Rebel Code: Linux and the Open Source Revolution*, London: Penguin.

Moniz, A. (2003) Work organisation in industry: practices of use of IT in Portugal, in Bechmann, G., Krings, B. and Rader, R. (eds) *Across the Divide: Work Exclusion: Work, Organization and Social Exclusion in the European Information Society*, Berlin, edition-sigma, pp. 197–216.

Morgan, G. (1997) *Images of Organization*, London: Sage.

Negri, A. (2003) *Time for Revolution*, London: Continuum.

Nietzsche, F. (1967) *The Will to Power*, New York: Vintage.

Nietzsche, F. (1969) *Thus Spoke Zarathustra: A Book for Everyone and No One*, London: Penguin.

Nietzsche, F. (1980) *On the Advantage and Disadvantage of History for Life*, Indianapolis: Hackett Publishing.

Nietzsche, F. (1990a) *Twilight of the Idols/The Anti-Christ*, London: Penguin.

Nietzsche, F. (1990b) *Beyond Good and Evil: Prelude to a Philosophy of the Future*, London: Penguin.

Nietzsche, F. (1994) *On the Genealogy of Morality*, Cambridge: Cambridge University Press.

Nonaka, I. (1991) The knowledge creating company, *Harvard Business Review*, 69, pp. 96–104.

Nonaka, I. (1994) A dynamic theory of organizational knowledge creation, *Organization Science*, 5, pp. 14–37.

Nonaka, I. and Takeuchi, H. (1995) *The Knowledge Creating Company: How Japanese Companies Create the Dynamics of Innovation*, Oxford: Oxford University Press.

Orwell, G. (1948) *Nineteen Eighty-Four*, London: Penguin.

Orwell, G. (1994) Politics and the English Language, in *George Orwell: Essays*, London: Penguin, pp. 348–359.

Ostergard, R., Tubin, M. and Altman, J. (2001) Stealing from the past: globalisation, strategic formation and the use of indigenous intellectual property in the biotechnology industry, *Third World Quarterly*, 22, pp. 643–656.

Parisi, L. and Terranova, T. (2000) Heat death: emergence and control in genetic engineering and artificial life, *CTheory*, ctheory.net/text_file?pick127.

Perelman, M. (1998) *Class Warfare in the Information Age*, London: Macmillan Press.

Piganiol, P. (1992) The emergence of corporate intelligence, in Sigurdson J. and Tagerud, Y. (eds) (1992) *The Intelligent Corporation: The Privatization of Intelligence*, London: Taylor Graham, pp. 23–38.

Plant, S. (2000) Information war in the age of dangerous substances, *The Journal of Cognitive Liberties*, 2, pp. 23–43.

Polanyi, M. (1967) *The Tacit Dimension*, London: Routledge and Kegan Paul.

Polster, C. (2001) How the law works: exploring the implications of emerging intellectual property regimes for knowledge, economy and society, *Current Sociology*, 49, pp. 85–100.

Porter, K. (2000) Terror and emancipation: the disciplinarity and mythology of computers, *Cultural Critique*, 44, pp. 43–83.

Porter, M. (1980) *Competitive Strategy: Techniques for Analyzing Industries and Competitors*, New York: The Free Press.

Porter, M. (1985) *Competitive Advantage: Creating and Sustaining Superior Performance*, New York: Free Press.

Porter, M. (1990) *The Competitive Advantage of Nations*, New York: The Free Press.

Porter, M. (2001) Strategy and the internet, *Harvard Business Review*, March, pp. 63–78.

Poster, M. (1995) Postmodern virtualities, in Featherstone, M. and Burrows, R. (ed.) *Cyberspace, Cyberbodies, Cyberpunk*, London: Sage, pp. 79–96.

Powers, T. (2002) *Intelligence Wars: American Secret History from Hitler to al-Qaeda*, New York: New York Review Books.

Quist, D. and Chapela, I. (2001) Transgenic DNA introgressed into traditional maize landraces in Oaxca, Mexico, in *Nature*, 414, pp. 541–543.

Rabinow, P. (1996a) *Essays on the Anthropology of Reason*, Chichester: Princeton University Press.

Rabinow, P. (1996b) *Making PCR: A Story of Biotechnology*, London: The University of Chicago Press.

Rashid, A. (2001) *Taliban: The Story of the Afghan Warlords*, London: Pan Books.

Reinerman, H. (2003) Internet portals in public administration. The readjustment of information and of administrative practice, in Bechmann, G., Krings, B. and Rader, R. (eds) *Across the Divide: Work Exclusion: Work, Organization and Social Exclusion in the European Information Society*, Berlin, edition-sigma, pp. 241–252.

Richardson, J. (2000) Sexual harassment and the legal procedure, in Hassard, J., Holliday, R. and Willmott, H. (eds) *Body and Organization*, London: Sage, pp. 215–229.

Rona, T. (1976) *Weapon Systems and Information War*, Seattle: Boeing Aerospace.

Ronfeldt, D. and Martinez, A.(1997) A Comment on the Zapatista 'Netwar', in Arquilla, J. and Ronfeldt, D. (eds) *In Athena's Camp: Preparing for Conflict in the Information age*, Washington: RAND National Defense Research Institute, pp. 369–391.

Rosen, S. (1993) *The Quarrel Between Philosophy and Poetry: Studies in Ancient Thought*, London: Routledge.

Sellars, J. (1998) Nomadic wisdom: Herodotus and the Scythians, *Nomadic Trajectories: PLI Warwick Journal of Philosophy*, 7, pp. 69–92.

Sigurdson, J. (1992) Introduction, in Sigurdson and Tagerud (eds.): 5–13.

Sigurdson J. and Tagerud, Y. (eds) (1992) *The Intelligent Corporation: The Privatization of Intelligence*, London: Taylor Graham.

Slow Food Society (1998) *Slow Food Manifesto, International Statute,* Online. Available http://www.Slowfood.com/.

Smith, W. (1963) *Mechanisms of Virus Infection*, London: Academic Press.

Stiglitz, J. (2002) *Globalization and its Discontents*, London: Penguin.

Taussig, M. (1999) *Defacement: Public Secrecy and the Labor of the Negative*, California: Stanford University Press.

Thomas, T. (1996) Russian views on information-based warfare, *Airpower Journal*, Summer, pp. 25–35.

Thompson, E. (1967) Time, work-discipline, and industrial capitalism, *Past and Present*, 38, pp. 56–97.

Toffler, A. and Toffler, H. (1993) *War and Anti-War: Making Sense of Today's Global Chaos*, New York: Little, Brown and Company.

Toffler, A. and Toffler, H. (1997) *Forward: The New Intangibles*, in Arquilla, J. and Ronfeldt, D. (eds) *In Athena's Camp: Preparing for Conflict in the Information age*, Washington: RAND National Defense Research Institute, pp. xii–xxiv.

Townley, B. (1994) *Reframing Human Resource Management: Power, Ethics and the Subject at Work*, London: Sage.

Tsoukas, H. and Vladimirou, E. (2001) What is organizational knowledge? *Journal of Management Studies*, 38, pp. 973–993.

Tsunetomo, Y. (2000) *The Hagakure: The Book of the Samurai*, New York: Kodansha.

Tuchman, B. (1984) *The March of Folly: From Troy to Vietnam*, London: Abacus.

Virilio, P. (1986) *Speed and Politics: an Essay on Dromology*, New York: Semiotext(e).

Virilio, P. (1990) *Popular Defense and Ecological Struggles*, New York: Semiotext(e).

Virilio, P. (1995) *The Art of the Motor*, Minneapolis: University of Minnesota Press.

Virilio, P. (1997a) *Pure War*, New York: Semiotext(e).

Virilio, P. (1997b) *Open Sky*, London: Verso.

Virilio, P. (1999) *Politics of the Very Worst*, New York: Semiotext(e).

Virilio, P. (2000a) *Strategy of Deception*, London: Verso.

Virilio, P. (2000b) *The Information Bomb*, London: Verso.

Virilio, P. (2002) *Desert Screen: War at the Speed of Light*, London: Athlone.

Webster, F. (1995) *Theories of the Information Society*, Routledge: London.

Webster, J. (2003) The value of women and men: gendering knowledge and skills in the information society, in Bechmann, G., Krings, B. and Rader, R. (eds) *Across the Divide: Work Exclusion: Work, Organization and Social Exclusion in the European Information Society*, Berlin, edition-sigma, pp. 101–116.

Weick, K. (1994) Cartographic myths in organisations, in Tsoukas, H. (ed.) *New Thinking in Organizational Behaviour*, Oxford: Butterworth-Heinemann, pp. 211–220.

Wiener, N. (1950) *The Human Use of Human Beings: Cybernetics and Society*, Boston: Houghton Mifflin.

Xerox (2001) *Industrial Business Matters 2001 Survey*, UK: Xerox Ltd.

Zuboff, S. (1988) *In the Age of the Smart Machine: the Future of Work and Power*, Oxford: Heinemann.

# Index